Building Space

David A. Dietzler

Copyright-©-2021-David A. Dietzler-All-Rights-Reserved

Cover painting by Dennis Martin

Table of Contents

Preface 4

Step One: The Basics 5

To Orbit and Back 11

From the Ground Up 21

Space Liners & Centrifuges 37

Large Ships 49

Power for Space Travel 60

Constructing Spaceships 65

Bootstrapping for Spaceships 68

Spaceships and Health 72

Pulsar Navigation 75

Collision Avoidance in Space 76

Interlunar Rocket Safety Mechanisms 79

Spaceship Accommodations 85

Space Docks 88

Attitude Control and Maneuvering 93

Money, Money, Money 95

Beyond the Moon and Mars 107

Spaceship Telecommunications 117

Work in the Future   Artificially intelligent robots are essential for the settlement of outer space and the creation of a space faring civilization beyond the bounds of Earth. Many people are afraid of this. Where will they work in the future? Will society have any use for them? Will the tech giants' promise of Universal Basic Income be kept? Will there be retraining for better jobs? Let's hope so! 119

Appendix 1: Mass of Shield 125

Appendix 2: Silane and Hydrogen Conservation 129

Appendix 3: Moon Ice 131

Appendix 4: Rocket Motor Cooling 132

References 134

Preface

All *Star Trek* fans know that the Enterprise is as big a star of the show as any of the human and Vulcan characters. Spaceships are cool, that's all there is to it. People want space travel. They want to experience weightlessness and standing on another planet while taking in scenic vistas. Our manned and unmanned space probes have not landed in the most interesting places, but there can be no doubt that there are sights never seen by human or mechanical eyes just waiting to be discovered on other worlds. The mountains of the Moon, the Valley of the Mariner on Mars, the skies of Titan and much, much more beckon to us. We humans are descended from hunting and gathering nomads. Wandering and exploring is in our nature. Like most primates we are a curious species. The Earth just isn't enough for us. We want to explore and settle outer space and that means we must have spaceships.

I propose several designs for interplanetary ships and ways to launch them into space, fuel them and provision them. Hopefully, this will pique your imagination and scientific spirit. Spaceships and space bases for mining, propellant production and manufacturing will cost billions of dollars, but there are people living today with enough money to invest in these dreams. Space travel will not be cheap, but it should someday be affordable to the common woman or man of Earth.

Step One: The Basics

Building spaceships-where do we start? How do we define "spaceship?" Do we mean vehicles that can travel from a planetary surface to orbit and return as well as ships that travel only from one planetary orbit to another and never touch ground? These are two entirely different kinds of vehicles but they definitely travel in space. Rockets with capsules and spaceplanes that can take off from Earth and reach orbit will be completely different from Moon landers like the Apollo LEM and Mars landers. The term lander is not accurate except for one-way vehicles. Since we want machines that can land and lift off again the terms "Moon Shuttle" or "Mars Shuttle" might be better. Ships that travel only in space and are made as low mass as possible for the sake of speed and efficiency are another entirely different animal. While space travelers might spend a few hours in a space capsule or a Mars Shuttle in which they only need about as much room as in an airplane, they could spend days, weeks even months in a ship that travels only in space. Therefore, such ships must be larger and carry more provisions of food, water and oxygen as do Shuttles. They must include spaces for sleeping, dining, showering and recreation. Robotic spaceships will be much different from manned spaceships. They won't need any amenities for human life so their design will be simpler and possibly more efficient. Cargo might be transported by robotic freighters that have no pressurized or radiation shielded cabins to cut costs. People make space travel more complicated and more expensive, but also more interesting.

Basics of Rocketry

A basic understanding of rocketry is helpful. The whole field of rocketry is dominated by the tyranny of the rocket equation. The initial mass of a rocket (M_i) when it is fueled and loaded divided by the final mass of the rocket (M_f) after it has expelled all its propellant or reaction mass is called the mass ratio. The natural logarithm of the mass ratio times the exhaust velocity of the expelled mass equals the final velocity or "delta V" of the rocket.

$V = c \log_e (M_i/M_f)$ or $(M_i/M_f) = e^{(v/c)}$ c = exhaust velocity v = rocket velocity

The higher the mass ratio (Mi/Mf) and the higher the exhaust velocity the faster the rocket. At a mass ratio of 2.718 to 1 the rocket will travel at a speed equal to its exhaust velocity. To go twice as fast a mass ratio of 7.4 to 1 is needed and to go three times as fast a MR of 20 is needed. Increases in speed require huge increases in propellant or reaction mass loads. That's the tyranny of the rocket equation. It gets worse. To increase the propellant mass larger fuel tanks are needed and that increases the mass final and requires a reduction of payload mass. A larger rocket carrying more propellant is heavier so it needs more thrust from its engines to get off the ground. To make matters even worse the rocket will endure air friction and gravitational deceleration as it climbs up from the Earth to LEO. When these aerodynamic and gravitational losses are accounted for a rocket needs to reach a theoretical speed of 9 km/sec to 10 km/sec rather than the orbital speed of 7.8 km/sec to reach LEO.

The bright side is that v is directly related to c. If the exhaust velocity is doubled the delta velocity is doubled. Rocket engineers have achieved high exhaust velocities with propellants like hydrogen and oxygen and they have built rockets with low mass fuel tanks using thin sheets of metal. They have also developed high thrust rocket engines. Staging is also used. Liquid fueled chemical rockets because of their high thrusts are just fine for reaching LEO; however, at least two stages are required and once in orbit the rockets' fuel and oxidizer is gone. If you want to go to the Moon or another planet in the solar system you need at least three stages. If it was possible to refuel in orbit the upper stage could travel to other worlds, but how do you refuel? Propellant could be launched up from Earth at great cost. Better yet, propellant could come from the Moon or near Earth objects (NEOs). That would require a large initial investment in Moon and/or NEO mining infrastructure. This would probably be worth it for the future of mankind and any other Earthly life forms that we might take with us into space.

Chemical rockets have reached the practical limit of exhaust velocity with liquid hydrogen (LH_2) and liquid oxygen (LOX) propellants. Nuclear rockets with LH_2 for reaction mass could go faster, but nuclear rockets present problems. Electrical propulsion like ion drive or VASIMR (Variable Specific Impulse Magnetoplasmic Rocket) offers much higher exhaust velocities but much lower thrust. Electrical propulsion can't get off the ground but it can accelerate spacecraft slowly in outer space to very high speeds. Energy can come from solar or nuclear sources. Since electric drives offer such high exhaust velocities either the rocket can reach much higher speeds with a given mass ratio or it can have a much lower mass ratio and much lower reaction mass load than a chemical rocket for the same delta V. If your reaction mass is coming up from Earth at great cost or it is coming from the Moon or a NEO you can save money by using electric propulsion and a smaller load of reaction mass. The only problem is that low thrust means long slow accelerations therefore it might take months to reach lunar orbit when chemical propulsion could get you there in days. This won't matter too much with unmanned vehicles. Electric propulsion can't land on the Moon or most other bodies in the solar system because of its low thrust so chemically propelled landers (Shuttles) are still needed.

The way to do things seems to be the use of chemical and electrical propulsion together. If the Moon and/or NEOs can supply relatively inexpensive propellant, spacecraft with high thrust chemical boosters could leave LEO and run up to escape velocity then activate electric drives and accelerate slowly for days even weeks and reach high speeds. This hybrid spacecraft could reach Mars or any other destination in the solar system faster than chemical or electrical propulsion alone. The infrastructure for building SSPSs could also supply propellant in the form of hydrogen, oxygen and combustible metal powders for deep space voyaging.

VASIMR is interesting because it can vary its specific impulse. Specific impulse abbreviated Isp is a rating of a rocket's efficiency. It is directly related to exhaust velocity. The Isp times the acceleration of gravity, 9.8 m/s^2 gives the exhaust velocity, c in the rocket equation, in meters per second. So a rocket with an Isp of 400 seconds has twice the exhaust velocity of a rocket with an Isp of 200 seconds and will go twice as fast with the same mass ratio.

Thrust is directly related to mass flow rate through the engine. If you can double the amount of propellant mass through the engine per second you can double the thrust. Thrust is directly related to Isp. Double the Isp and you double the thrust. With electric drives you have high Isp but low mass flow rates, so you get low thrusts. The power required for a rocket engine goes up to the square of the Isp. Double the Isp and you need four times as much power. This would double the thrust, but if you keep thrust the same or one half that which you get with the doubled Isp by reducing the mass flow, you only need twice as much power. The Isp is directly related to the power required for the same thrust.[1] Keep thrust constant by reducing mass flow and doubling the Isp only doubles the power requirement.

With VASIMR, it is possible to get a higher thrust at a lower Isp and accelerate to escape velocity. As the rocket speeds up it can lower its thrust by decreasing mass flow through the engine and increase its Isp by adding more power directly with increase in Isp. In this way it can "shift gears." Ultimately, with the higher exhaust velocity or Isp it can reach high speeds and get to Mars in about 39 days.[2] This would require a lightweight high power reactor. Vapor core reactors might do the job.

Another way to look at it is this. In low gear you trade away engine RPM and horsepower for torque. In high gear you trade away torque for RPM or speed. With VASIMR at low Isp, we trade away "fuel efficiency" in terms of kilograms of propellant per delta V in meters per second, for thrust to climb out of a gravity well. At high Isp we trade away thrust for fuel efficiency and this reduces propellant requirements and allows more delta V.

In conclusion, resources from the Moon and eventually NEOs, combinations of chemical and electric propulsion, lightweight nuclear and solar power systems and low cost reusable rockets to LEO can open the doors to the solar system. Lunar and asteroid resources can also supply more than propellant. They can supply oxygen, water and food once space farms are established.

Life Support Basics

Rockets of all kinds can get us into space. Life support systems (LSS) will make it possible for us to stay in space for long periods of time. Early LSS for spaceships and habitat will use supplies of oxygen and chemical CO_2 absorbents. It might also be possible to freeze CO_2 and other pollutants out of the interior air with cryogenic systems. Extra oxygen could be extracted from lunar rocks and regolith. Eventually, space farms and green plants will recycle the CO_2 and generate oxygen. This is called Closed

Ecological Life Support Systems or CELSS. We can expect CELSS for space stations and surface bases after some sustained development. It seems doubtful that spaceships will have CELSS due to the large amount of space crops will need. However, we might eventually see huge interplanetary and even interstellar space arks that can travel in space for years, even decades, with CELSS.

Water can be recycled by condensing humidity from the air and filtering, distilling, purifying and sterilizing waste water from sinks, showers, toilets, etc. Mechanical systems can do this until CELSS is established in space stations and surface bases and biological systems for recycling water are in action. Ships will use mechanical water recycling. Food is mostly water so dehydrated and freeze-dried foods will be rehydrated with the same water over and over again. Whole food would be very bulky and water supplies to sustain crews for months at a time would be far too massive and expensive to ship into space.

Ultimately, it will make more sense to grow food in space instead of ship it into space even if it is dried out. Space farms must be established early on at manned bases so that the first harvests can be made within a few months. Fresh food will be much better for morale than dried foods. Crops will require controlled temperatures. So will humans. Habitat will require thermal insulation plus heating and cooling systems. Cosmic ray shielding will also be necessary. This would probably consist of several meters of regolith to cover the habitat. That would also serve as a barrier to micrometeorites.

Ships will depend on farms in space stations and surface bases to get food. Wastes from ships will be recycled by the space farms and CELSS of space stations. Ships will require thermal insulation in the form of lightweight polyurethane barriers and heating and cooling systems. Cosmic ray shielding will be far too massive for ships. About seven tons of polyethylene or water would be needed per square meter of hull area to get the cosmic ray dose down to 20 mSv/yr.[3] That's too much mass. One way to get the cosmic ray dose down is to spend less time in space by using high speed nuclear electric propulsion systems. Even then, a substantial amount of mass shielding will be needed to protect crews from solar flare and coronal mass ejection radiations. These radiations are less energetic than cosmic rays but they are still dangerous. Ships will need solar flare shelters with several hundred kilograms of water or polyethylene per square meter. More advanced ships might have superconducting magnetic radiation shields that deflect the charged particle swarms of cosmic rays and storms on the Sun. Magnetic shields should be lighter than mass shields therefore suitable for ships that make long voyages in space, but no experiments have been done with spacecraft, so there isn't much hard data to guide our designs. It is known that exposure to strong magnetic fields like those from an MRI scanner can cause people to experience vertigo, flashes of light and a metallic taste in the mouth.[4] The biological effects of prolonged exposure to powerful magnetic fields is unknown. If this proves to be unhealthy, it may be possible to shield humans from the magnetic field with mu metal, permalloy or Metglas, presuming this wouldn't add excessive mass to the ship. Micrometeor impacts are rare and some kind of self sealing hull system might be used. Collision with larger objects will be avoided with radar.

Weightlessness can be detrimental to health. It can cause bone loss and muscular atrophy as well as other problems. Humans have spent long periods of time in zero-G and survived. Russian cosmonaut Valeri Polyakov spent almost 438 days in space. American astronaut Scott Kelly spent a year on the ISS.[5] The effects of low lunar and Mars gravity over long periods of time is unknown. Exposure to weightlessness can be reduced by faster ships and shorter travel times. Space stations can rotate and larger ships could have centrifuges to produce partial "gravity" at least. Perhaps some "artificial gravity" provided by centrifuges, even if less than one Gee, will be beneficial for space travelers, but no hard data exists, so we can only speculate.

To Orbit and Back

Pure spaceships that travel from Earth orbit to lunar orbit, a Lagrange point station, Mars orbit or elsewhere in the solar system, that never touch the ground or enter an atmosphere and never endure high Gee forces can be built very lightly. They don't need high thrust engines to climb out of a gravity well, but with less mass they will use less propellant or reach higher velocities and achieve shorter transit times. Such vessels will need plenty of room for supplies, life support systems, living, sleeping and recreating. Humans will spend days, weeks, even months aboard such ships. Naturally, these ships won't be much good if there is no way to reach orbit and return to Earth or land on another world. There must be rockets that can climb up to orbit and return operating on Earth, the Moon, Mars and other destinations. These orbital vehicles will be designed specifically for work on their particular world and there must be refueling infrastructure. They must be reusable to make space travel affordable.

SpaceX Falcon rockets are at work today. They are partially reusable and fairly economical. The Dragon capsule is fully reusable and has completed several missions to the ISS. The SpaceX Super Heavy and Starship have yet to be demonstrated. There is no way to predict the future for the British Skylon spaceplane, Blue Origin rockets or other possible rockets. Rockets that can reach low Earth orbit need 1.5 to 2 stages, but many have dreamed of single stage to orbit vehicles (SSTOs). To return to Earth they need heat shields. Moon Shuttles can reach lunar orbit or a Lagrange point with just one stage because the Moon's gravity is so much weaker than Earth's. To land, they don't need heat shields and parachutes or wings because the Moon has no atmosphere. They don't even need to be aerodynamic at all because they will work only in a vacuum. However, without an atmosphere to brake in they need more propellant to retro-rocket down from space. Vehicles similar to Moon Shuttles could also operate on asteroids, small airless moons, dwarf planets and the planet Mercury. Mars Shuttles can reach orbit with just one stage thanks to the low gravity of the red planet. They could burn hydrogen and oxygen from Martian water or methane and oxygen from water and carbon from the atmosphere. Since Mars has a thin atmosphere, Mars Shuttles will need heat shields, but they can also aerobrake and use parachutes, thereby reducing the propellant requirements for retro-rockets. Titan Shuttles would be interesting. The gravity is low, somewhere between the gravity of the Moon and Mars, and the atmosphere is thicker than Earth's. Titan Shuttles might even use nuclear ram-jets to climb into orbit and they could parachute or glide down to soft landings.

I suggest the development of a rocket based on a modified Space Shuttle External tank, liquid fueled booster, reusable engine module, passenger capsule and payload faring/cargo container. This rocket is similar to the F-1 Flyback equipped rocket proposed in the seventies that was projected to orbit 400,000 pounds of payload. [6] The booster could be fueled with kerosene and LOX or methane and LOX. It would help propel the rocket to an altitude of about 20 miles then separate and navigate its way back down with hypersonic air foils then fire its rockets briefly to make a pin-point landing on a barge at sea like Falcon rockets do today. If it can't be accurate enough, perhaps it would land in a large net made of retro-rocket exhaust resisting tungsten steel. The liquid hydrogen and LOX burning reusable main engine module would operate from sea level to LEO. For maximum efficiency it could be an aerospike engine that works at any pressure level. A whole new engine would have to be developed, since no aerospikes are at work today. The engine module would have a heat shield and parachutes. After one orbit around the Earth it would detach from the ET and payload, fire its retros and re-enter the atmosphere. It would parachute down and eject the disposable heat shield just seconds before touch down and inflate air bags for a soft landing. The module would be recovered from its landing field and somehow be transported to a factory where it is refurbished and prepared for another flight. It would be fairly large so transport by truck, train or even the Super Guppy aircraft might be impossible. Barge transport might be the only option.

The payload and external tank would be propelled all the way to orbit. The tank would be used in space as a source of metal for construction, so there would be no waste. To reduce costs the design of the external tank must be simplified and it must be mass produced on a mostly automated assembly line. Payload farings and cargo containers would also be used in space.

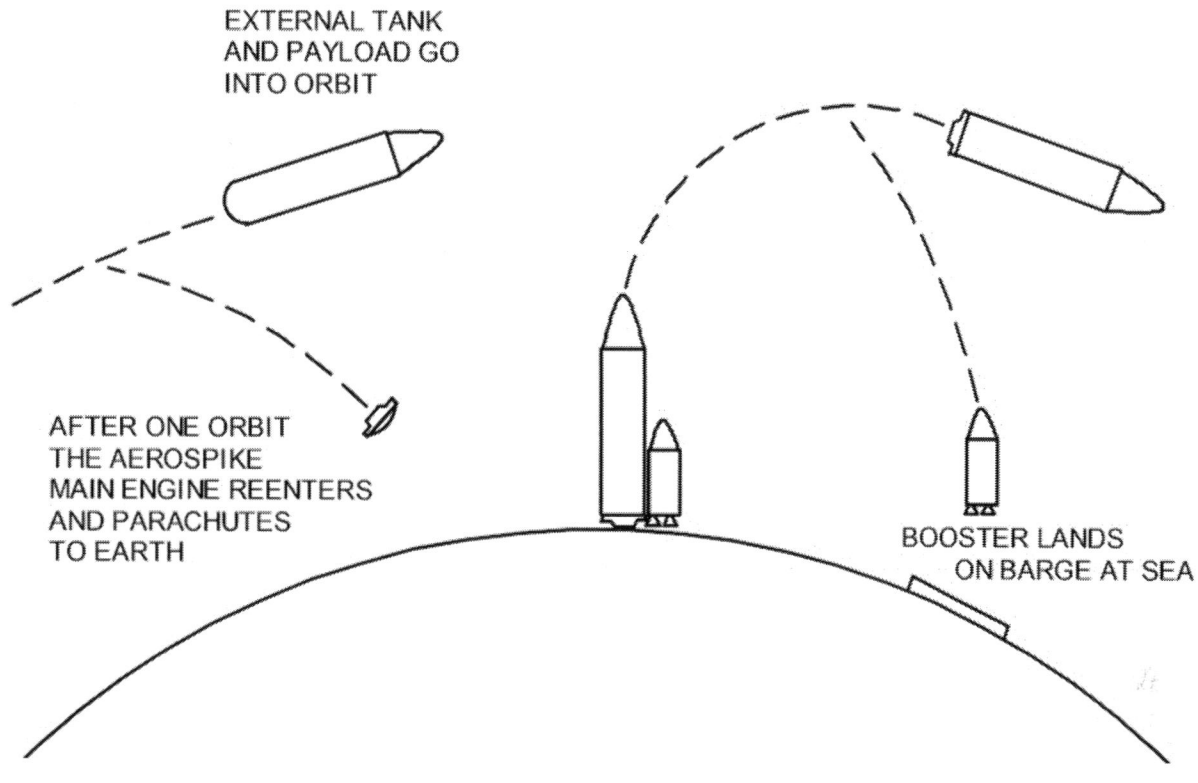

Fig. 1 Reusable rocket system

The ET weighs about 60,000 pounds. It consists of aluminum and titanium alloys with a coating of polyurethane. All these materials would be valuable in Earth orbit. Lots of materials could be sourced from the Moon, but probably not copper, lithium or vanadium found in the alloys that compose the ET, unless there are minerals on the Moon that we don't know about. The external tank adds 30 english tons of metal to the actual payload and this could be used for construction and manufacturing in space. The tanks could be cut up with electric saws or lasers. Various machines launched up to LEO or built from lunar materials in space could shape the metal plates. The metals could be melted and cast, powdered or drawn into wires and fed into 3D printers to make all sorts of things. The tanks as they are might be fitted with airlocks and interior furnishings and used for space station or spaceship modules. They could be fitted with pipes, pumps, solar shields and other parts to make orbital propellant depots. The alloys would probably be too valuable to powder and burn as rocket fuel.

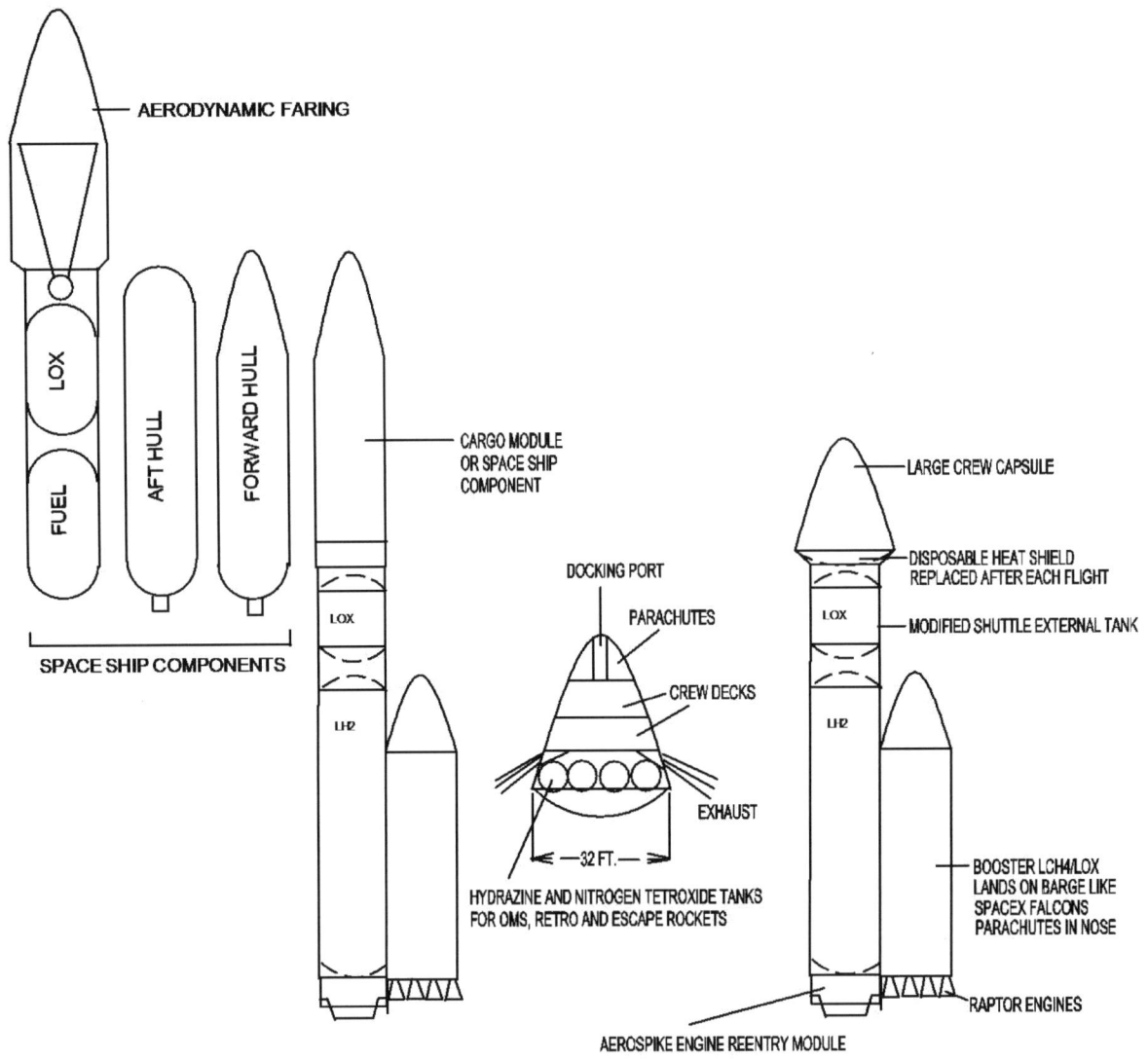

Fig. 2 Space Shuttle ET based rocket configurations.

The rocket could orbit spaceship and space station sections that are assembled in space. Passengers could be transported in a large capsule. If they are really packed in there might be 100 passengers in the capsule. If they are given a more generous 3 ft. by 3 ft of space about 60 passengers could be carried into space and returned. The capsule could use hypergolic propellants which can be stored for long periods of time in space for orbital maneuvering, attitude control thrusters and retro-rockets. The capsule would deploy a series of parachutes to slow down after re-entry and fire its retro and orbital maneuvering system rockets briefly for a soft touch down on land. If the rockets fail, the heat shield could be ejected and air bags inflated to absorb some of the energy of impact and keep passengers from being killed, although they would experience quite a jolt. This calls into question the wisdom of using hypergolic propellants that ignite upon contact with each other. What if the fuel and oxidizer tanks rupture after a hard landing? There could be a deadly explosion. The tanks could be surrounded by shock absorbing materials or they could use a non-hypergolic combination of soft cryogens like liquid methane and LOX that are kept in super insulated tanks to prevent boil off.

Conditions in the capsule would be pretty tight, like a jet airliner, but passengers wouldn't have to spend a lot of time onboard. It would only take about ten minutes to rocket up from an equatorial launch base to a space station in equatorial low Earth orbit (ELEO). Launching from Russia or KSC to the ISS often involves several hours of orbital maneuvering to dock with the station because their orbits are highly inclined to the equator. Ideally, launch bases would be located on or very near the equator and space stations would be in ELEO. This also avoids the radiation of the South Atlantic Anomaly. Launch windows to or from a station in ELEO occur about every 100 minutes, so scheduling is simplified with equatorial launch bases and return to Earth in case of an emergency is expedited.

The Atlantic coast in Brazil, the Indian Ocean coast in Somalia and the Pacific coast in Borneo could all serve as equatorial locations for launch bases. Airports, hotels, hospitals, warehouses, stores, shipping ports, power plants, factories and launch base facilities including propellant manufacturing systems would have to be developed in these remote locations. An international corporate effort rather than a government program with all kinds of national security concerns could do the job. As for the cost of getting there by air, anyone who couldn't afford to fly to Brazil, Somalia or Borneo couldn't afford to travel in space.

If the capsule's orbital maneuvering system (OMS) and retro-rocket engines have enough thrust, they could be used as escape rockets to propel the capsule away from a malfunctioning or even exploding rocket beneath during ascent. The capsule would have to be watertight and capable of landing at sea although normally it would touch down on land.

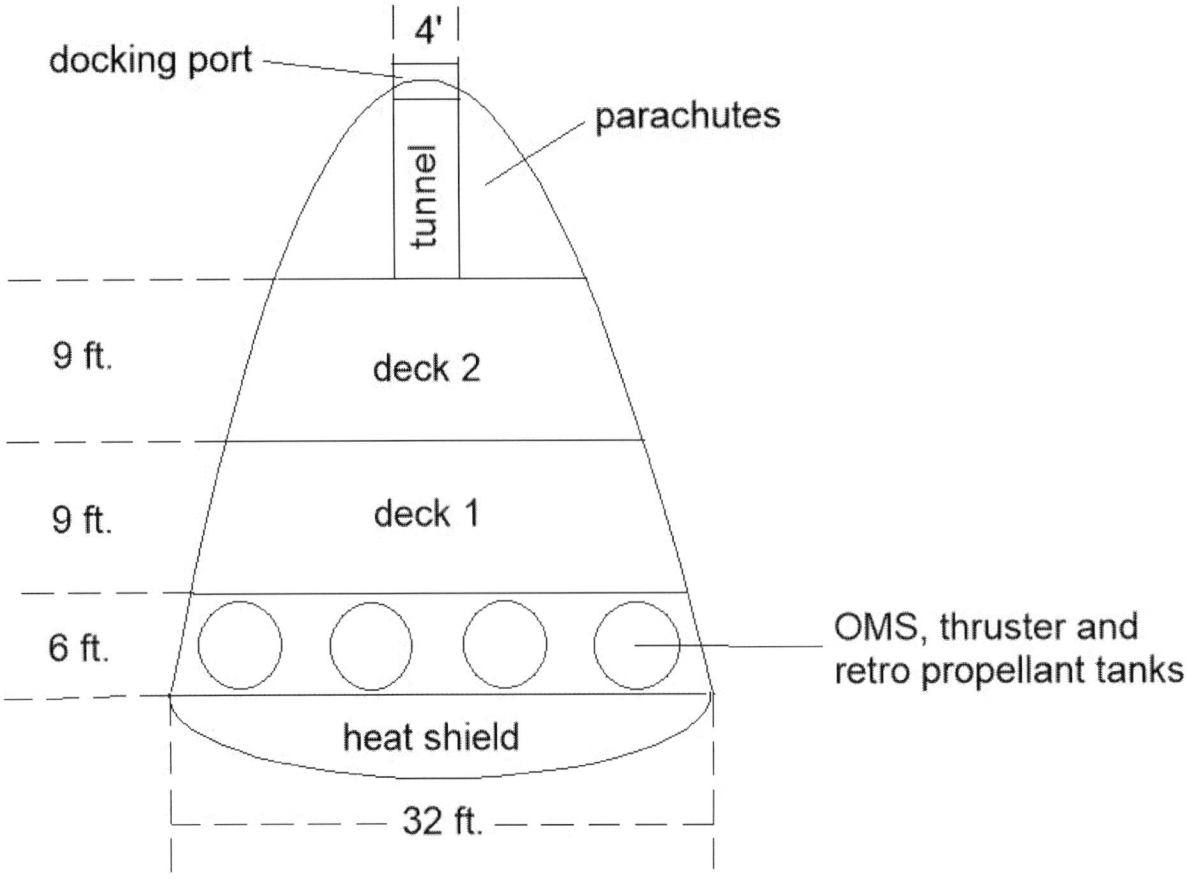

Fig. 3 Large passenger space capsule.

The capsule might have windows, but it isn't really designed for orbital sight-seeing. It would dock at a space station where there would be plenty of time to watch the Earth below. Space stations would be constructed from inflatables and hard modules rocketed up from the ground as well as retro-fitted external tanks. If there is enough traffic to ELEO there would be no shortage of ET materials for building space hotels and eventually large Kalpana stations, although that might also require materials from the Moon. Orbital propellant depots stocked with lunar LOX, sometimes called LUNOX, and metallic powder fuels would make it possible to go beyond ELEO and visit the Moon, Mars and beyond.

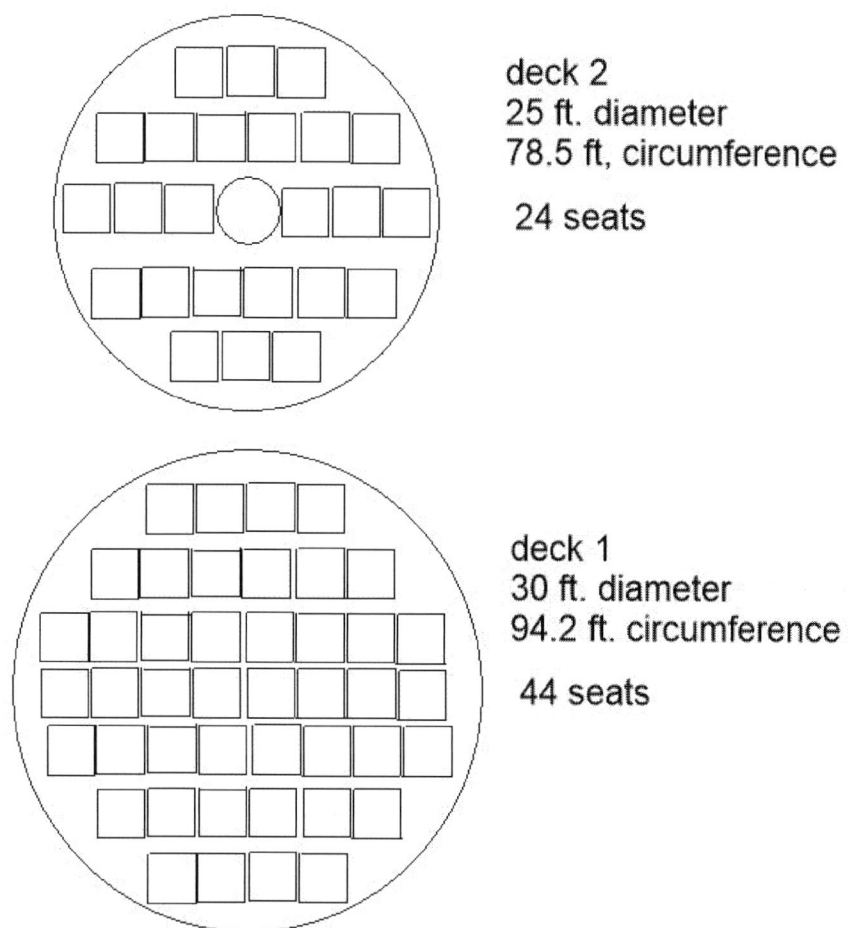

Fig. 4 Space capsule decks with seats approximately 3' x 3' for 68 passengers. Some seats might be for crew including pilots and flight attendants. Complete automation is also possible.

The creation of a space tourism industry without space industrialization for other purposes like helium 3 mining or space solar power satellite (SSPS) construction is kind of "iffy." No government would spend taxpayer dollars for expensive space vacationing without causing an enormous public uproar. Even helium 3 mining and SSPS construction will be challenged by radical environmentalists who sometimes place emotionalism and superstition over scientific reason. With space industrialization and materials from the Moon and later from near Earth asteroids it should be possible to create orbital and lunar resorts.

Lunar tourism means there would have to be pure spaceships for travel from ELEO to low lunar orbit (LLO) or Earth-Moon Lagrange point 1 (EML1). There would also have to be Moon Shuttles. Lunar regolith is 40% oxygen by mass, so there will be no shortage of LUNOX. Powdered metals might serve as fuels. There has already been some experimentation with monopropellants consisting of aluminum powder suspended in liquid oxygen.[7] If hydrogen can be extracted from polar ices at a reasonable price, it could be used for rocket fuel, but that seems like a waste. There could be carbon compounds in the ice, but using them for liquid methane rocket fuel would be an even bigger waste when carbon is needed for CELSS and synthetic materials. Hydrogen could be conserved if it is combined with abundant silicon which composes about 21% of the regolith to make silane. A rocket burning silane, SiH_4, with LOX having a specific impulse of 340 seconds will use more propellant overall but only half as much hydrogen as a LH_2/LOX rocket with an Isp of 450 seconds and only one fifth as much hydrogen as a nuclear thermal rocket with an Isp of 1000 seconds.[8]

Hydrogen supplies could be extended even further if powdered metals are suspended in a silane carrier fluid to be burned with LOX in bi-propellant rockets. Silicon and aluminum both burn with about 13,000 BTU/lb. Magnesium, calcium and iron also burn in pure oxygen but with less heat. Magnesium/LOX gels are said to be shock and vibration sensitive and tend to detonate. Magnesium suspended in silane shouldn't present a problem. There isn't much hard data on silane/metal powder rocket fuels and there is room for lots of research into this matter. Hydrogen is needed to make water, synthetic materials and chemicals. It shouldn't be wasted. The cost of mining lunar polar ice and launching it into space is unknown. Once the ice is obtained extracting hydrogen is just a matter of melting the ice in solar heated containers and putting it through electrolysis. It might be cost effective to send LH_2 up from Earth with passenger rockets. Several tens of tons of LH_2 might be "piggybacked" with passenger flights since the payload capacity of the rocket described above will probably be much higher than the mass of the capsule. Silane would be made in Earth orbit with silicon from the Moon along with LUNOX and metal powders to make propellants.

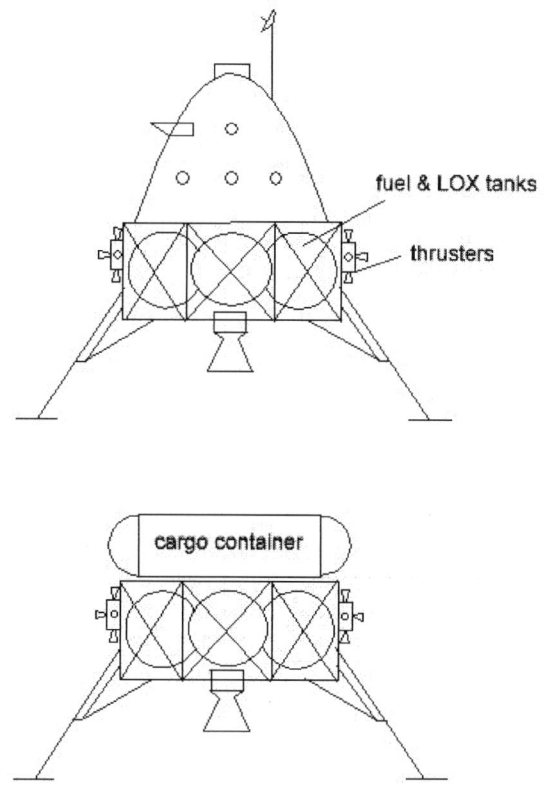

MOON SHUTTLES

Fig. 5 Rocket vehicles for moving passengers or cargo between the lunar surface and lunar orbit.

Shuttles that work on the Moon and other airless worlds don't need to be aerodynamic at all. Shuttles that work on Mars or Titan will need aerodynamic design. They could be shaped like capsules with heat shields on the bottom side. Landing legs could be extended at touchdown. During flight they would fold up inside the vehicle like landing gear on aircraft. Mars has plenty of water in its regolith and carbon in its atmosphere and polar ice caps. Producing hydrogen or methane and LOX on Mars will not be too troublesome. Raw CO_2 could be pumped down from the atmosphere, compressed, cooled and liquified and used as working fluid in nuclear thermal rocket motors. On Titan it would be possible to extract methane and ethane from the atmosphere and from hydrocarbon lakes to be used with nuclear thermal rocket motor propelled Shuttles.

From the Ground Up

In the early decades of space industrialization and tourism, spaceships of all sorts will be built on the Earth and rocketed up to orbit in parts that will be assembled in space. Spaceships will be complex vehicles with elaborate wiring and piping systems. At least the software will weigh next to nothing. It won't be possible to build spaceships from scratch on the Moon or in space until a high degree of manufacturing sophistication is reached off-Earth.

The first things that come to mind when designing a spaceship are the hulls and propellant tanks. The external tank might be a good place to start. A lithium-aluminum alloy ET has a mass of only 58,500 lbs. and measures 27.5 ft, X 154 ft. A hull of similar dimensions could be developed. It would not have to be separated into fuel and LOX tanks like the ET. It could be made on a largely automated assembly line to cut costs. The interior of the hull would be fitted with a forced air hull thermal equalization system, decks made of sheet metal and plastic, seats, toilets and waste handling systems, air vents, fans and CO_2 scrubbers, oxygen and water tanks, cabin heating and cooling systems, batteries, electronic controls, lighting systems and miles of wiring and many other parts. Small armies of engineers will have to work for years to design ships and gangs of technicians will be needed to assemble them. The sections of the ships would be mounted atop a rocket like the one described in the previous chapter and launched into space where they are assembled at a space dock. Then they will have to undergo years of rigorous testing to make sure they are safe. Even if a Federal Space Travel Administration doesn't exist to enforce safety regulations the companies that build the ships will want to make sure they are safe so they don't get a bunch of people killed and wind up getting sued into bankruptcy by the families of the dead space farers. The cost of designing, building and testing spaceships will run into the tens of billions of dollars. Maybe more.

A preliminary set of designs for a ship capable of traveling from Earth orbit to lunar orbit and back is pictured below along with some of its sub-systems. It is based on Shuttle ETs.

CHEMICALLY PROPELLED INTERLUNAR PASSENGER SHIP 30 HRS. LEO TO EML1 OR LLO

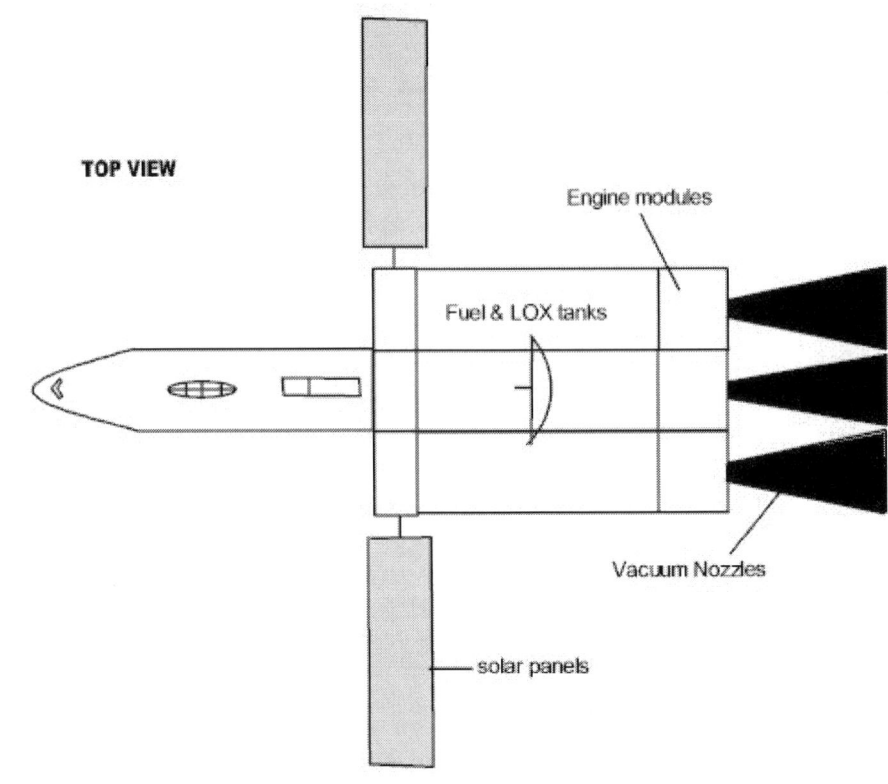

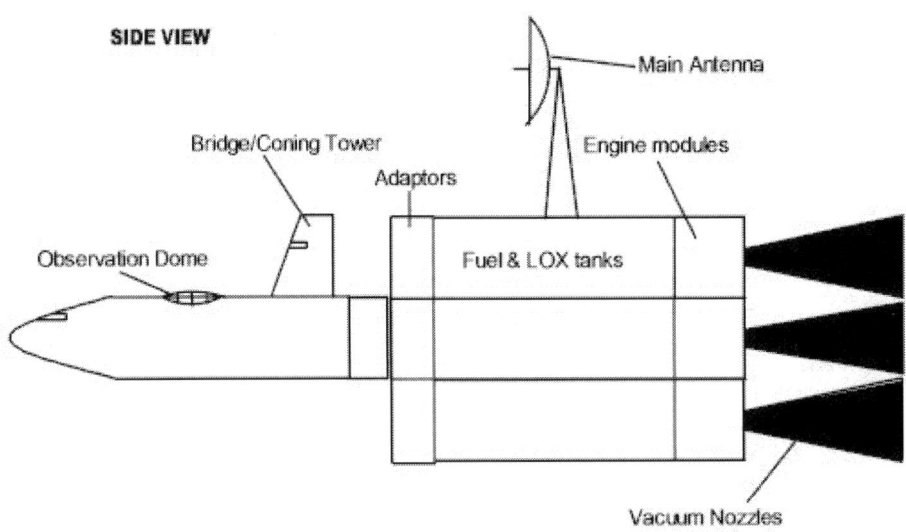

Fig. 6 Ship composed of six hull sections based on SSET.

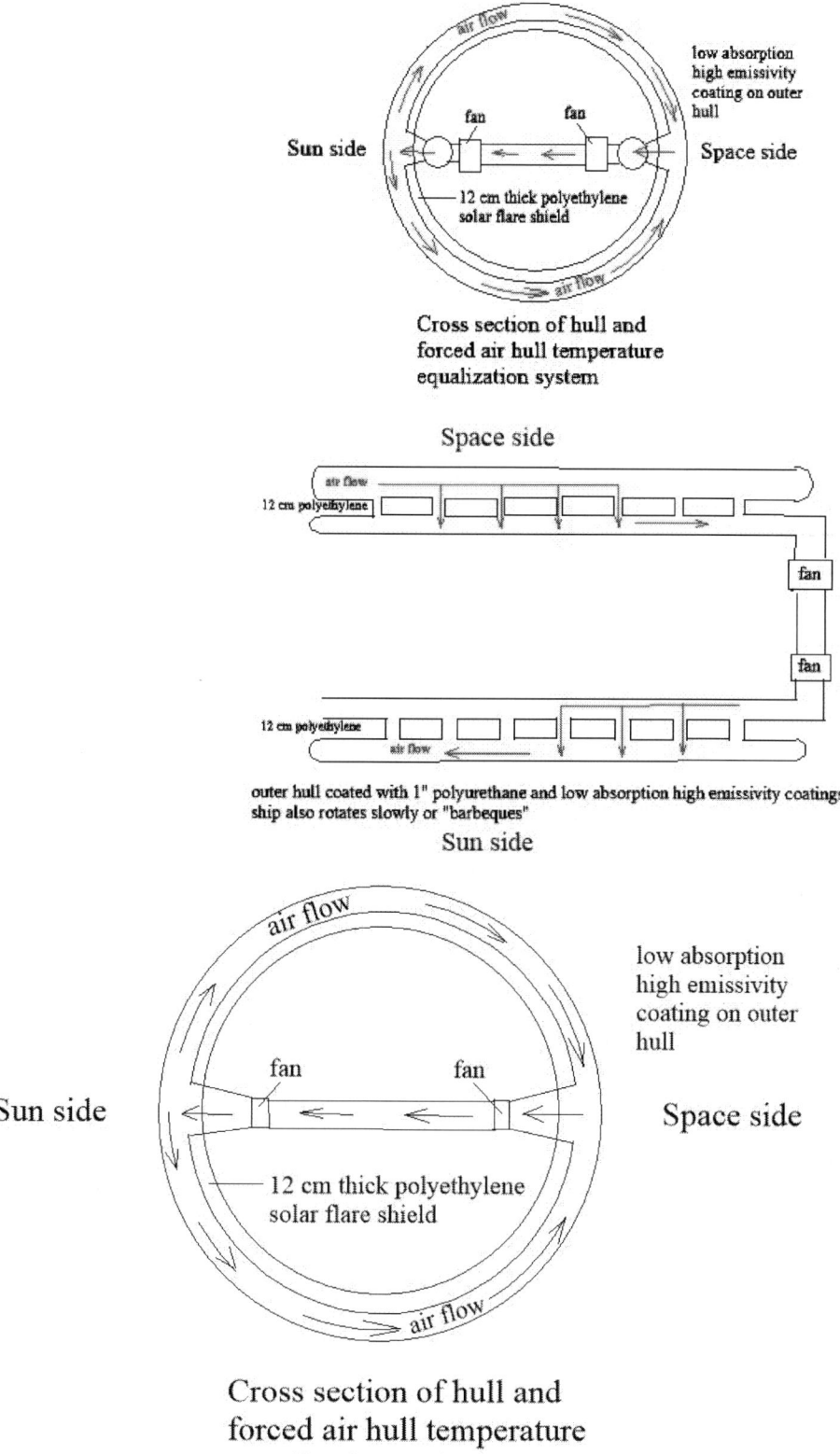

Fig. 7 Forced air temperature equalization. No coolant leaks.

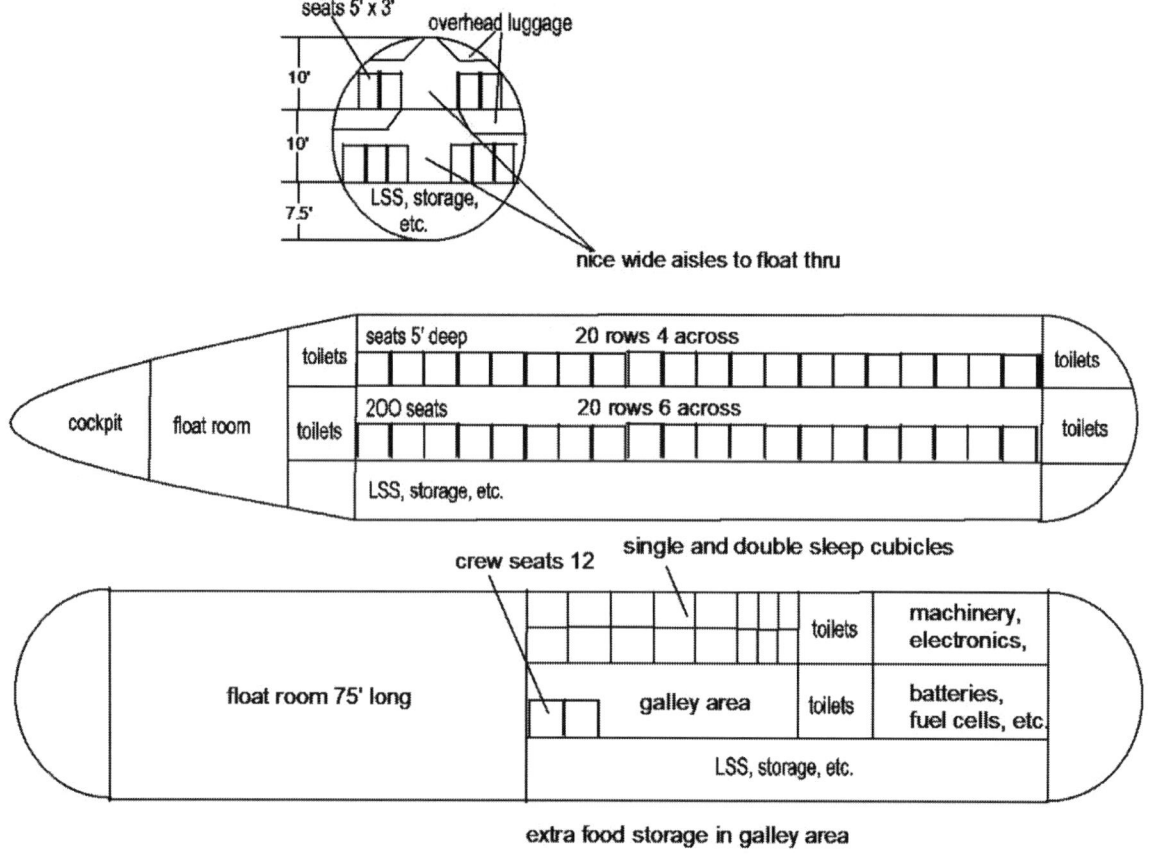

Fig. 8 Upper: Cross section and lateral section of forward hull Lower: Lateral section of aft or rear hull.

The ship will take only 30 hours to reach the Moon. Escape velocity from LEO is about 11 km./sec. At that speed the ship could reach the Moon in about 42 hours. If the ship speeds up to 11.2 km./sec. at trans-lunar injection it can cut that down to 30 hours. [9] That still seems like a long time to be crammed into a seat but the seats will be larger than those provided on a jet airliner and everybody will be weightless, so there won't be pressure on their backs and lower bodies. There will be curtains to draw around seats for privacy and sleeping, and for those who don't like that arrangement there will be sleeping cubicles in the rear hull. There will also be spacious floating rooms where passengers can enjoy weightlessness and stretch their limbs. There will be numerous air suction toilets for bodily necessities. Seems like a nice spaceship, but what's wrong with this picture? The answer: Radiation.

Using NASA's On-Line Tool for the Assessment of Radiation In Space (available at oltaris.nasa.gov), it can be determined that the galactic cosmic ray dose (GCR) at 1 A.U. from the Sun in 30 hours is just 2 or 3 mSv. That's about as much radiation as a person receives on Earth in a year, aside from medical exposure, and the legal radiation exposure limits for nuclear workers is an average of 20 mSv/yr. over five years with no more than 50 mSv in one year. Cosmic rays won't be a problem for short trips to the Moon, except for crews who spend a lot of time in space. Van Allen Belt radiations are nothing to worry about either. Solar flares and coronal mass ejections are the dragons we must fear. Most of the time there are no solar storms going on so most space travelers won't have to face this danger. When there are solar storms, or solar particle events (SPE) to use the technical term, bad things will happen. With no shielding a solar flare could dump out enough radiation to kill half the people exposed.

Whole Body Effective Dose Equivalents

November 1960 SPE

Shielding Thickness	Dose in millisieverts
20 g/cm^2	138
35 g/cm^2	66

August 1972

20 g/cm^2	60
35 g/cm^2	29

September 1989

20 g/cm^2	57
35 g/cm^2	35

AVERAGES

20 g/cm^2	85
35 g/cm^2	43

This data is based on a lithium-aluminum hull about 0.25 inches thick and polyethylene shields with 20g/cm^2 and 35 g/cm^2. Let's round off the averages for ease and caution to 100 mSv and 50 mSv. With 20 grams per square centimeter of hull area the solar flare dose can be reduced to about 100 mSv and with 35 gr/cm² it can be reduced to about 50 mSv.

One Sievert of radiation increases a person's chance of developing a fatal cancer by 5.5%.[10] Based on the linear no-threshold model a 100 mSv dose will increase the chance of getting a fatal cancer by 0.55% and a 50 mSv dose about 0.275%. The average American's chance of getting cancer is about 40% and the chance of dying from cancer is about 20%. It looks like about 3 out of a thousand space travelers exposed to a solar flare even with 35 g/cm² of shielding will die of cancer due to that exposure and their survivors will then sue for everything they can get. Will they sign a legal release? Or will people tolerate the dangers of space travel the way they accept the dangers of tobacco, alcohol, fatty foods, extreme sports and driving? We can only hope that medical science advances to the point at which cancer can be prevented by inoculation or prophylactic medicines and with future treatments that when cancer does develop it can be cured most of the time.

What about heavier shielding? That's difficult. I only allowed 12 g/cm² in the diagrams above based on comments by a popular space expert, but NASA's OLTARIS shows us that that is not enough. Maybe it's enough to prevent immediate death from a solar flare, but it is not enough to keep too many space tourists from dying of cancer years later and their survivors litigating for damages.

A polyethylene shield with 35 g/cm² would be about 15 inches thick. It would have a mass of about 350 kilograms per square meter of hull area. Perhaps we can build a solar flare shelter. The floating room in the rear hull could have this much shielding. The shield mass would be about 245 metric tons or about 540,000 lbs. Plastic tiles would have to be shipped into space and installed by workers. The rocket described previously can only lift 100 to 200 english tons to LEO so several flights would be necessary. The only problem is cramming 200 passengers and a dozen crew members into the solar flare shelter for hours. With the crew in the solar storm shelter the ship will have to run on autopilot or be teleoperated by space station crews.

This would be one hell of a voyage. Interlunar flight could be avoided during times of intense solar activity. Perhaps astronomers will learn to predict solar weather with great accuracy. Even so, some people are going to get blasted by a solar flare and cramming everyone into the rear hull's floating room will not be pleasant. The passenger seating area could be shielded instead. That would require a shield with a mass of about 315 metric tons. Crew areas could also be shielded so they can remain in control of the ship. It looks like we will be adding about 350 metric tons of shielding to the ship, and that's a lot, but why be cheapskates? A heavier ship will require more propellant and that means higher costs, but those will have to be passed on to the customer. Paying more for an already very expensive trip to avoid the discomfort of being awoken in the middle of your sleep and herded into a solar flare shelter will probably be acceptable.

So how much propellant do we need? The forward and rear hulls will total about 60 tons. I imagined four ET sized tanks for the rockets. That would be another 120 tons. The passengers would have a mass of about 20 tons and all the seats and machinery might add another 60 tons. So that's 260 tons and to that we add 350 tons of solar flare shielding for a total of 610 tons. This thing is turning into the Battlestar Galactica!

In LEO we are going 7.8 km/sec and we want to increase that to 11.2 km/sec. That requires a 3.4 km/sec burn. Then we have to slow down to enter lunar orbit. Since we left Earth at more than escape velocity we will be entering the Moon's zone of gravitational influence at about 2.1 km/sec, maybe faster, even though we only exceeded escape velocity by 0.2 km/sec. This is found by the equation: *hyperbolic excess speed squared equals burnout speed squared minus escape velocity squared or:* $V^2_\infty = V^2_{bo} - V^2_{esc}$ or $V^2_\infty + V^2_{esc} = V^2_{bo}$ If you want to know more about this study up on the Oberth effect and hyperbolic velocity. Since an object falling from infinity will reach escape velocity, it seems the Moon's gravity will accelerate us by another 2.4 km/sec. However, in the same way that a small excess of velocity over V_{esc} leads to a larger excess of velocity at "infinity," the converse is true. It will take less reduction in velocity to be captured into orbit. Our speed at periapsis will be $2.1^2 + 2.4^2 = 3.19^2$ Since we need to slow down to lunar orbital speed of 1.6 km/sec we need to lose 1.59 km/sec. Our total delta V will be around 3.4 + 1.59 = 4.99 km/sec and we will round that up to 5 km/sec.

Now we turn to the rocket equation.

We are burning silane and LOX at 340 seconds. That's 3.332 km/sec.

$e^{(5/3.332)} = 4.48 = MR$ $610+X/610 = 4.48$ $X = 2123$ metric tons of propellant.

Silane and LOX burn with a 1:2 mixture by mass. Thus, 708 tons will be silane and 1415 will be LOX. Since liquid silane at -112° C., according to https://encyclopedia.airliquide.com/silane, has a density of about 0.58 g/cc, we need about 1220 cubic meters of SiH_4 tank volume. LOX has about 1.141 g/cc so we need 1240 cubic meters of LOX tank volume. Total tank volume is 2460 cubic meters. An ET has a total volume of about 2050 cubic meters-550 for the LOX and 1500 for the LH_2. The propellant tanks for the silane/LOX rocket will be different. Instead of four ETs with a total mass of 120 tons it looks like we can use smaller tanks and save some mass. Four dual tanks with 615 cubic meters each would be cylinders about 30 meters long and 5.11 meters wide. For comparison, an ET is about 47 meters long and 8.3 meters wide. Since we are working with ballpark figures and estimates let's guess that these tanks would weigh 20 metric tons. That cuts the ship's mass down by 40 tons.

610-40 = 570 tons. 570+X/570 = 4.48 MR X = 1984 tons of propellant.

Now we need 1140 cubic meters for silane and 1159 for LOX. Our tanks could be a little smaller and lighter, and our total propellant load less.

Unless you like math, I am sure I lost many of you several paragraphs ago. For those who stuck with me, there is plenty of research and calculation left to do. I am not the man with all the answers. If you can find the answers you often get paid real money.

So now we have to redesign our spaceship with smaller propellant tanks. That won't cause anyone any pain. If there is a SPE the passengers won't have to leave their seats or give up toilet access. They won't have to cram into a storm shelter with only a few cubic meters of volume per person. The propellant tanks might provide some radiation shielding for the rear hull, but people sleeping in cubicles will have to move back to their seats in the forward hull. We can't please everyone.

Of course, a price must be paid in the form of propellant because of the mass of the shield. If we need 1984 metric tons of the stuff, and we can get it from the Moon for $4/kg. then we spend $7,936,000 or $39,680 per passenger. Maybe we can get the price down to $3/kg. and spend $5,952,000 or $29,760/passenger. That's just for a one-way flight. For a round trip propellant alone will cost $59,520 to $79,360 per passenger. Perhaps the cost of propellant can be reduced. Since it will come mostly from the Moon, prices could be lowered after the mining base is paid for.

Earlier writings in previous books painted a rosier picture. I imagined round trip flight to the Moon at a price of $100,000. Now it looks like that will be pricier. It's always possible that the companies will just worry about breaking even for transportation and make their real money from hotel fares, food, booze, gambling and spectator events on the Moon.

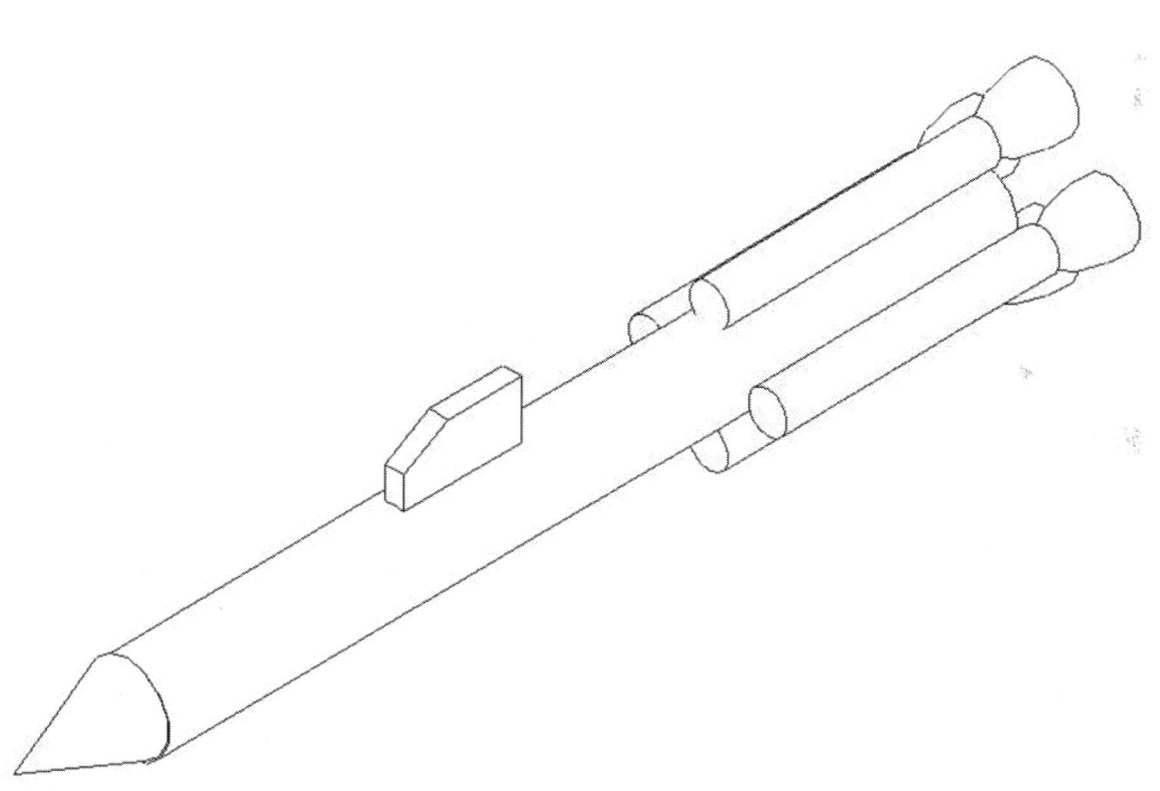

Fig. 9 3D CAD by Mark Rode. Note smaller propellant tanks.

If a silane and metal powder slurry fuel is used the fuel might be denser and less voluminous. When silane burns it forms silica and water. If a slurry is used metal oxides will form too. Deposits will probably form in the rocket engine combustion chamber that have to be removed periodically with ultrasonic probes perhaps. This won't work with a turbine to drive fuel and LOX pumps. Deposits on the turbine blades and in the bearings would ruin it. If turbopumps are used they might use hydrogen and oxygen. A simpler way to do things is to use pressure feed. Gaseous silane could pressurize the fuel and gaseous oxygen could pressurize the LOX. These gases stored in titanium alloy high pressure tanks could also feed ullage motors needed to settle the propellants. The only problem with pressure feed is that either real heavy tanks to withstand the pressure are needed or the propellant flow and thrust must be lower with lighter tanks and lower pressures. That's not a big problem in space where you don't have to overcome gravity to lift off. Lower thrusts and lower rates of acceleration will work.

Turbines don't just run fuel and oxidizer pumps in a rocket engine. They also power hydraulic pumps that are part of the steering mechanism. Perhaps the engines could be swiveled with jack screws and electric motors. Power would come from lithium batteries that can be recharged by multi-junction solar panels. Since the engines will only fire for several minutes the jack screw motors won't use a lot of energy.

A ship from LEO to EML1 or LLO and back will have to fire rockets when it leaves LEO and again when it brakes into EML1 or LLO. It will refuel then rocket away from L1 and retro rocket back into LEO. The ship will have to fire rocket motors 4 times with each round trip. How many firings can it make before motors have to be replaced or overhauled?? I propose simple pressure fed rocket motors. Just valves for the pressurant tanks and valves into the firing chambers but no turbopumps. They will be very simple and thus very reliable, inexpensive and simple to replace or refurbish. Pressure fed rockets aren't used on Earth because the only way to get high fuel/oxidizer flow rates for high thrust that's needed to overcome gravity is to use very high pressures and that means the tanks have to be real heavy. In space you don't need much thrust to accelerate a ship so lower pressures and lighter tanks can do the job. On Earth you need more thrust than the weight of the rocket to get it to lift off. In space thrust equal to the mass of the rocket will produce acceleration at 1G. If thrust is only 1/4 as much as the mass of the ship it will accelerate at about 0.25 G which is enough to reach escape velocity in short notice. Since a ship in LEO needs to add 3.2 km/s to reach escape velocity and 0.25 G equals 2.45 m/s^2 it will take only 22 minutes to achieve escape velocity. During acceleration metal powders suspended in silane would tend to settle out towards the rear of the tank. To prevent this, an agitation system will be needed in the fuel tank. Ullage motors burning silane and oxygen gases will also be needed to move liquid propellants to the rear of tanks before firing main motors.

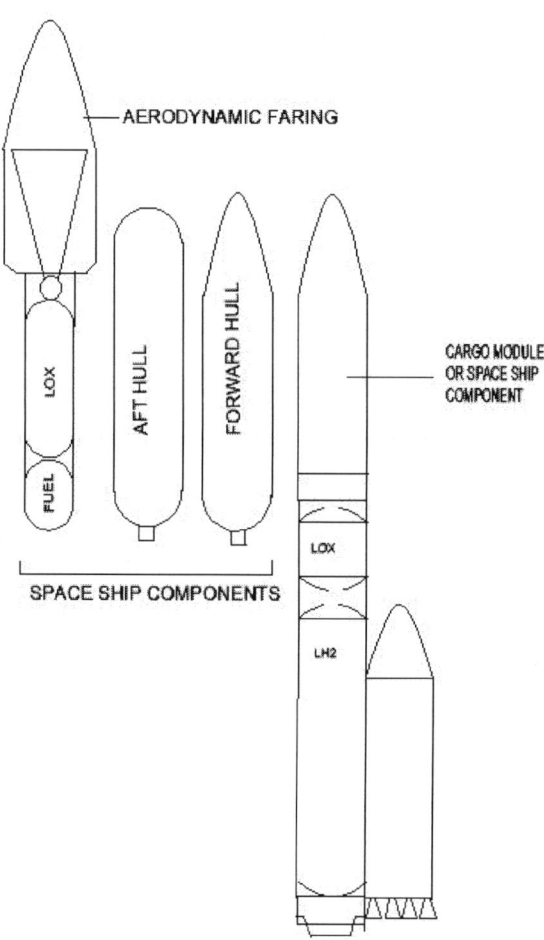

Fig. 10 Forward and aft hulls with smaller propellant tanks and rocket engine.

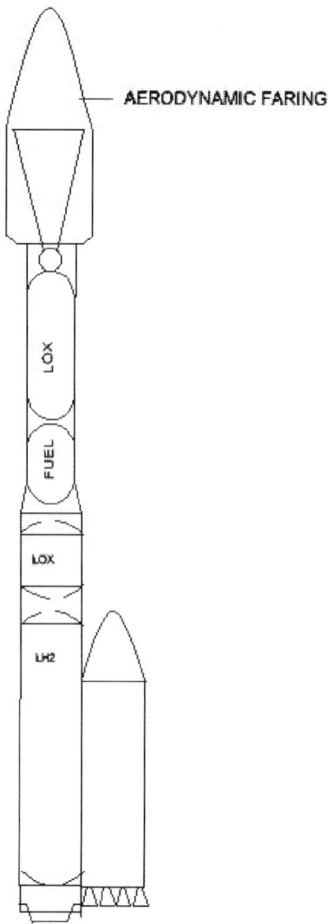

Fig. 11 Smaller propellant tanks and engine mounted atop rocket for transport to LEO space dock/assembly station.

Two hulls with interior and exterior fittings and four propellant tanks with engine modules will require six launches. Six external tanks or about 180 tons of aluminum and titanium alloys will also be orbited. The polyethylene tiles or bricks for the radiation shield might require three or four more launches. It seems there will be no shortage of external tanks in LEO if this system is employed.

Space stations consisting of scaffolding and robotic arms will be needed to assemble the hulls and tank+engine modules. A small crew could be stationed in space and the robotic arms could be teleoperated by ground crews via communication and tracking satellites in GEO. Decades of infrastructure development on the Moon and in space must precede the production of spaceships for lunar tourism.

There will have to be Moon mining bases and mass driver lunar launchers at work. Mass catchers to receive lunar materials and haul them through space will be needed. Powersat construction will have to be going on in GEO. This is how the first stage of space industrialization will make profits. The SSPS builders in GEO will sell their excess oxygen, silicon, iron and calcium to LEO propellant depots, production stations and space hotel builders. Since the SSPS builders will be smelting the lunar regolith to get the aluminum, magnesium, titanium and other elements they need they might charge more for the left-over oxygen, silicon, etc. than they paid the Moon miners for it. On the other hand, they might pay to have the stuff taken off their hands!! That would reduce propellant costs for lunar tourism.

There will be more to consider than just raw material costs for propellant. There will also be the cost of hydrogen piggybacked into LEO or from the Moon. To that will be added the cost of making silane from silicon and hydrogen. There will be operating costs and the amortized cost of the spaceships and orbital stations.

Spaceships could reach the Moon in 30 hours followed by 12 hours of inspection, maintenance including steam cleaning of the passenger cabin to prevent pandemics, restocking the galley, emptying the waste tanks and refueling with gaseous silane and oxygen for pressure feed, thrusters and ullage motors in addition to liquid silane and oxygen for the main engines. There will have to be space stations in LLO or at EML1 where all this can be done. The ships will then return to LEO with a new load of homebound passengers. That will also take 30 hours followed by 12 hours of refurbishment. Since there are 168 hours in a week, two round trips per week per ship are possible. If 100 round trips are made every year with two weeks for major overhauls, one ship could move 20,000 passengers per year. If they only pay $100,000 each then the ship could bring in two billion dollars per year, before costs. The only way a round-trip ticket is going to cost just $100,000 is if propellant is much cheaper than $3 or $4 per kilogram. There will be other costs including payroll for the crew and the maintenance technicians when the ship is docked.

Burning silane will cut hydrogen consumption in half compared to hydrogen/oxygen propulsion, but will it really be cheaper? Hydrogen could be piggybacked from Earth but there would only be a limited supply. Lunar polar ice could provide hydrogen, but mining in deep supercold craters will probably not be easy. That alone would require a polar base or several of them and a fleet of ice mining machines with either nuclear or beamed power. Then the ice has to be melted and the water electrolyzed into hydrogen and oxygen. The hydrogen could be liquified and stored in small tanks then transported to the mass driver base on the lunar equator and the tanks of LH_2 launched into space. It might also be possible to rocket liquid hydrogen into space. It seems as if all this will make hydrogen an expensive commodity. Ice will be in demand for water alone and hydrogen as well for the synthesis of organic chemicals for synthetic materials. The high demand and limited supply could drive the price of hydrogen way up.

There will be no shortage of oxygen and silicon. Lunar regolith is 40% oxygen and 21% silicon. These two elements are as plentiful as dirt. Extracting them from regolith is not difficult. They can be obtained anywhere on the Moon. At the equatorial mining base excavators will simply pit mine regolith, package it or sinter it into spheres, and launch it to EML2 with mass drivers. Mass catchers will receive the loads and haul them to stations in LLO or at EML1 and in GEO. In GEO the regolith will be smelted to get metals and the by-product oxygen, silicon, iron and calcium will be hauled down to LEO with solar electric tugs. Smelting to obtain oxygen and metals will go on in LLO or at EML1 also. Some elements will be used for propellant and some will be used to bootstrap space stations or "space ports" in LLO or EML1.

In LEO, iron could be converted to steel which only requires small amounts of carbon and this could be used to build space stations, in addition to all those external tanks and their valuable aluminum and titanium alloys. It might be possible to re-oxidize silicon for glass and calcium for lime to make mortar and cement. It might also be possible to burn powdered silicon in excess of that needed for silane, powdered iron and calcium as rocket fuels to stretch hydrogen supplies even farther.

Space stations in LEO will be protected from GCRs and SPEs by Earth's magnetic field. Stations in LLO or EML1 will need heavy shielding unless light weight magnetic shields are developed. A minimal amount of shielding will be needed for electronics. Humans will need to be shielded from GCRs as well as SPEs and that might require 7 tons of water or polyethylene per square meter of hull. If this proves to be impractical, the stations in cis-lunar space will have to be mostly automated, teleoperated and human crews would work in them only in brief shifts. Spaceship crews could stay in well shielded stations or ride down to surface bases with passengers. On the surface, habitat will be covered with 11 tons of regolith per square meter to protect people from GCRs and SPEs. Crews could take layovers of several days to give their bodies time to repair DNA damage done by space radiation. If they endure an SPE they might have to give up their space careers depending on how much radiation they have absorbed over the course of their duty in space.

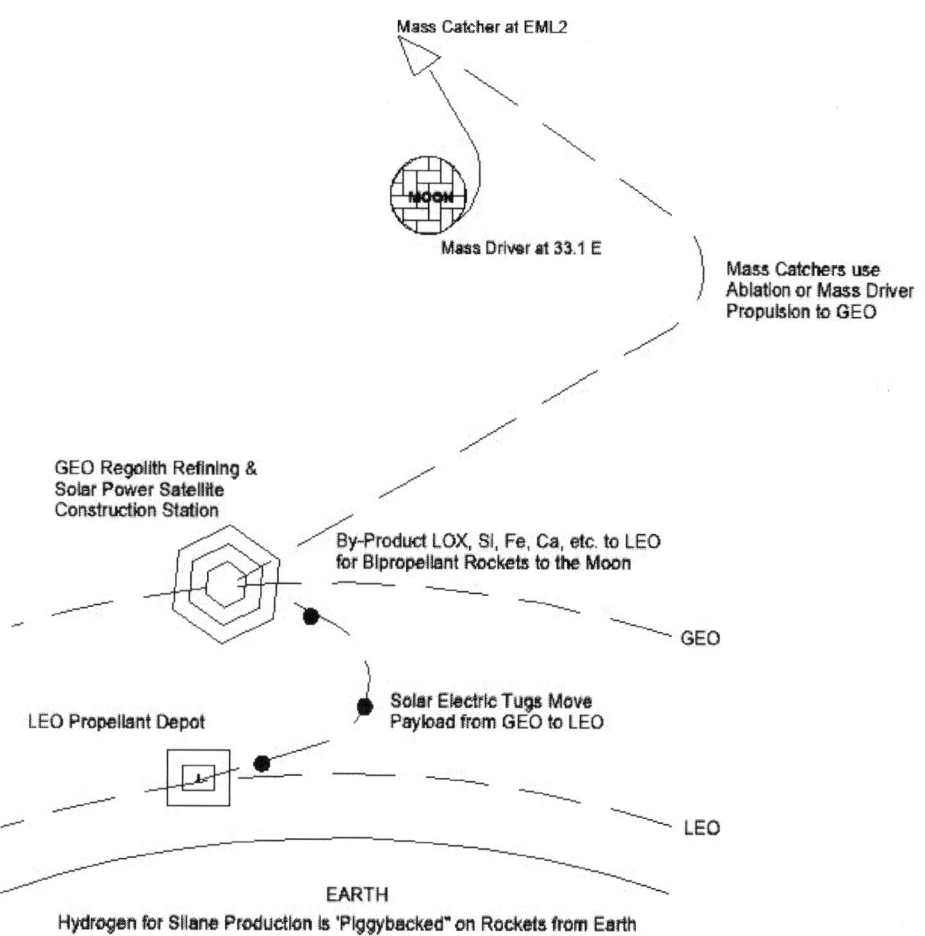

Fig. 12 Moving materials from the Moon to Earth orbit.

Space Liners & Centrifuges

Artificial gravity plating is about as unlikely in the near future as negative matter reactionless drives. The only way to provide a substitute for gravity is to use centrifuges. The first spaceships will be weightless. They will be arranged sort of like jet airliners, use chemical rocket propulsion and reach the Moon in about 30 hours. That might seem like a long time to sit in a couch even if the seat can recline and you can sleep. Since passengers will be weightless they won't get very sore sitting for a long time but they will want to move around and stretch their limbs. Weightless floating rooms with observation domes perhaps will be provided to make the flight more enjoyable and increase passenger comfort.

If the ship operates on the same time schedule as the launch base and everyone has adjusted to that time, lights could be turned down and everyone can sleep at the same time. If not, then there could be curtains around seats and if that doesn't help there could be sleeping cubicles as pictured in a previous chapter in the rear hull.

There will probably be people who dislike this arrangement, perhaps because they get really space sick or want more privacy, who will prefer to travel in private cabins in centrifuges that provide "artificial gravity." While complete spaceship hulls could be built on the ground and rocketed into orbit where they are docked together, centrifuges will have to be built in space from metals obtained from spent external tanks and possibly lunar materials. Space "shipyards" with ways and scaffolding to support pieces of the centrifuges as they are welded together will be required.

Ships with centrifuges will be later generation vehicles. These centrifuges will add mass to the ship so it might take more propellant to reach the Moon in such ships. That means the ticket price must be higher. If the ship uses less fuel and takes longer to reach the Moon, the ticket price must be higher because the ship won't make as many round trips per year. The amortized cost of the ship will be higher. Putting it another way, it will take longer to pay off loans and interest will be higher. Equipping the ships with centrifuges will make them more expensive also. Slower flights means more hours for crews and more payroll per flight; and more provisions. It might be more economical to use more fuel and make more flights. Larger numbers of passengers could travel on the ship and costs could be divided up. Perhaps 200 passengers will ride in the weightless section and another 200 in the centrifuges. Radiation shielding to protect everyone from SPEs will add lots of mass too.

The centrifuges will be about 90 feet in diameter and rotate at 3 RPM to produce just a little less than lunar gravity equivalent. The cabins will be about 14 feet by 14 feet including the private toilet closet. Water for flushing will be stored in tanks and pumped through lightweight plastic pipes to the low mass compact plastic toilets. The sewage will be pumped through plastic pipes and into an electric distillation machine in the centrifuge. The distilled water will be filtered, treated with chlorine or ozone and reused for flushing toilets.

The centrifuges will rotate in opposite directions to cancel out gyroscopic effects. Flywheels on the ship's roll axis might also be needed to control the ship's attitude as angular momentum changes when people move in and out of the centrifuges. There will be centrifuges with private cabins that are entered by ladder thru a sliding door in the ceiling instead of wasting space with wide corridors. There will also be centrifuges with a galley where meals can be cooked normally thanks to the "artificial gravity" and passengers can sit down for a meal, pour drinks, etc. There will also be exercise rooms and hot showers. The water for the showers might be recycled after distillation, filtration and chemical treatment.

Waste water from the kitchens (galleys), showers and toilets will be stored in tanks and off-loaded when the ship docks at a space station or space port. Fresh water will be taken on-board and waste water will go into the station's CELSS system to be processed into nutrient solution for hydroponic farms where crops are grown to feed space travelers and workers. To off-load the waste water and pipe in the fresh water through flexible plastic hoses put in place by maintenance crews, the rotation of the centrifuges will have to be stopped. This doesn't seem to be a problem until we consider the water in un-flushed toilets and drain goosenecks. Powerful pumps in the drain lines will have to draw or suck the water out of the toilets and drains and pump it into waste water storage tanks. Then the centrifuge's rotation can be stopped.

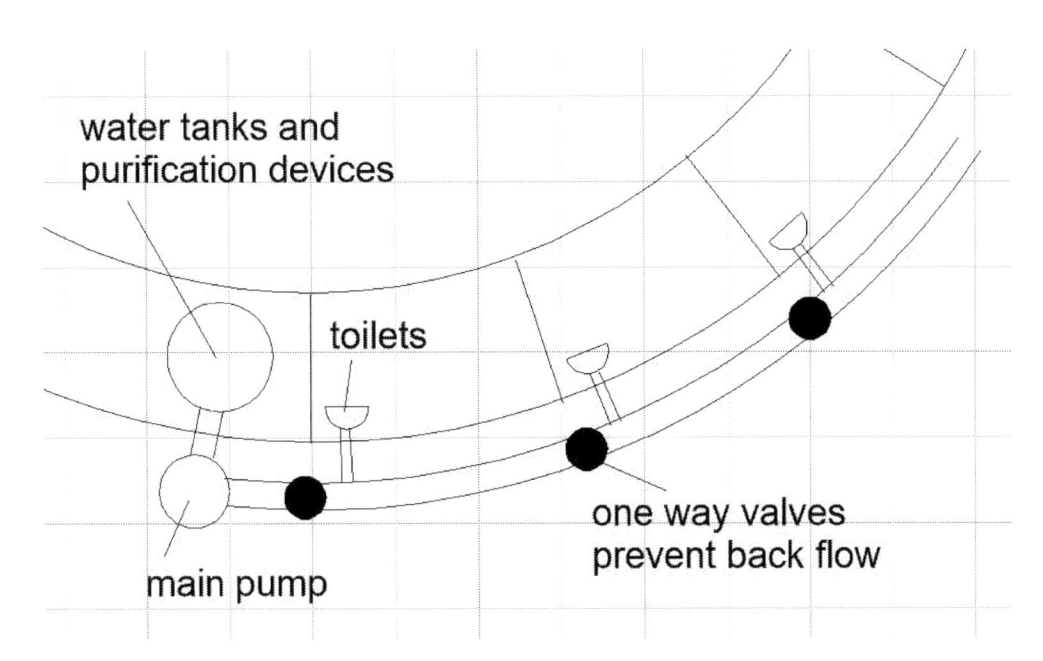

Fig. 13 Main pump draws water out of toilets with suction. After toilets and drains are cleared, rotation of centrifuge can be stopped and maintenance can be performed.

CHEMICALLY PROPELLED INTERLUNAR PASSENGER SHIP 72 HRS. LEO TO EML1 OR LLO

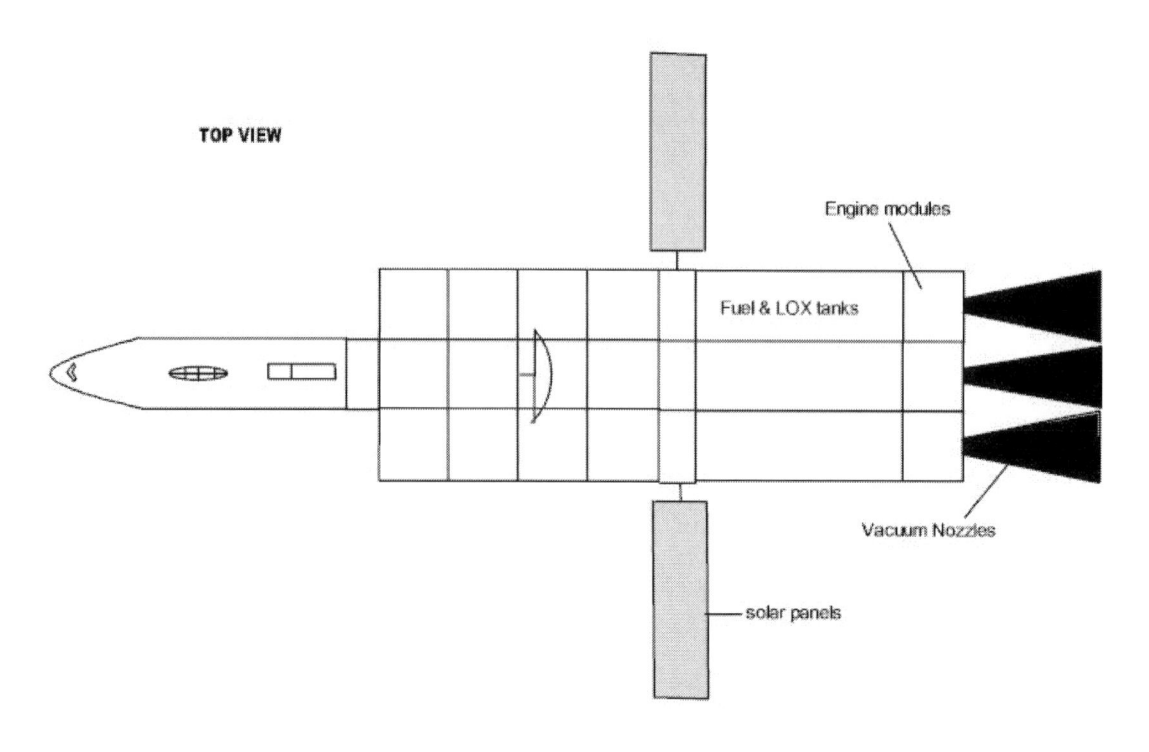

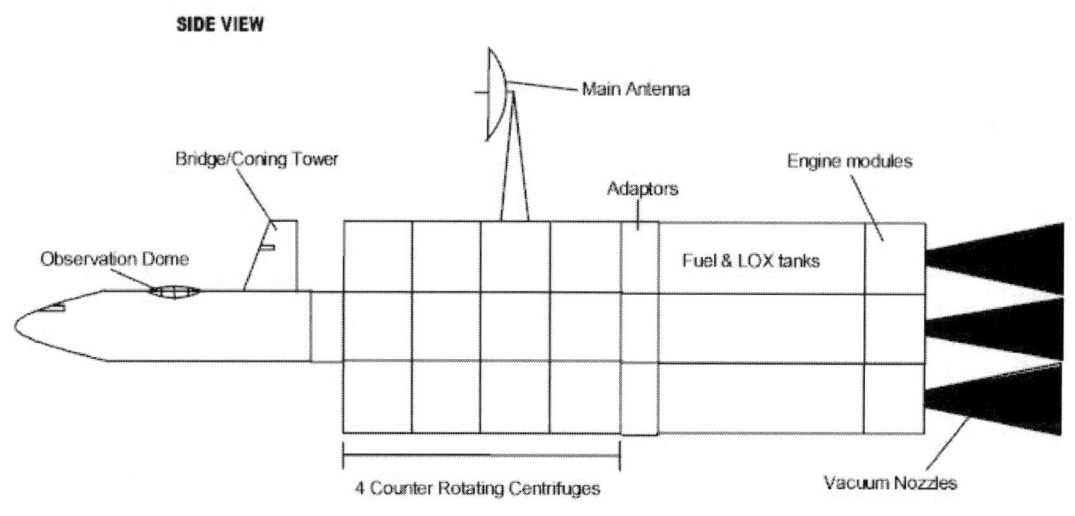

Fig. 14 Ship with centrifuges

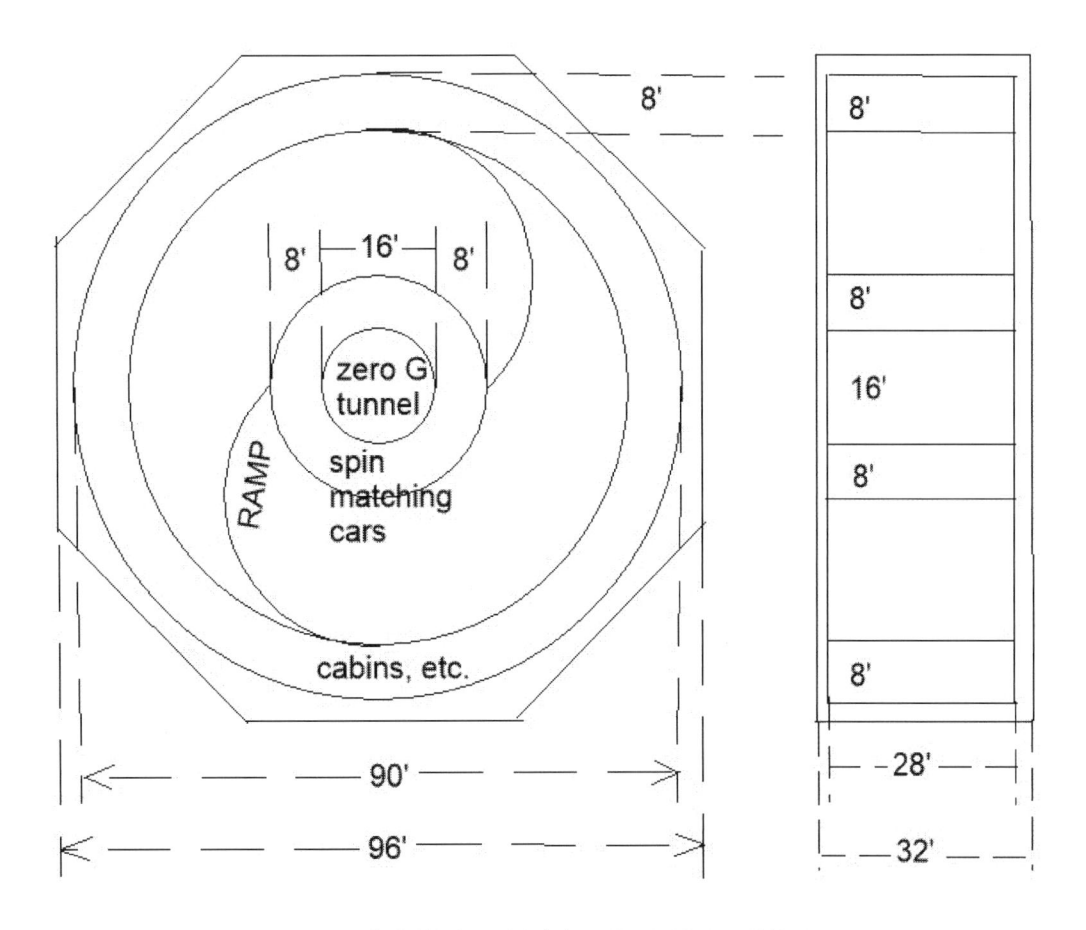

GENERAL CENTRIFUGE LAYOUT

Fig. 15 General layout with dimensions

Passengers will float through the central tunnel, possibly pulling themselves hand over hand along ropes. Getting around in weightlessness will require some minimal physical ability. They will then enter some stationary "elevator" cars that when filled with people will rotate until matching the spin of the centrifuge. Then they can walk down ramps to the cabins or mess hall in the rim.

Moving in and out of centrifuges, that is transferring from low G to weightlessness then low G again might make some people ill. Even with centrifuges it won't be possible to escape from weightlessness. Of course, weightlessness is the whole reason many people want to travel in space! Medications to control motion sickness may be necessary for some people. Low gravity will make life onboard more natural and more comfortable, but not much is known about the long term effects of low gravity. It is known that prolonged weightlessness is very unhealthy. Ships to Mars and other destinations in the solar system will need centrifuges.

MESS/GYM CENTRIFUGE LAYOUT

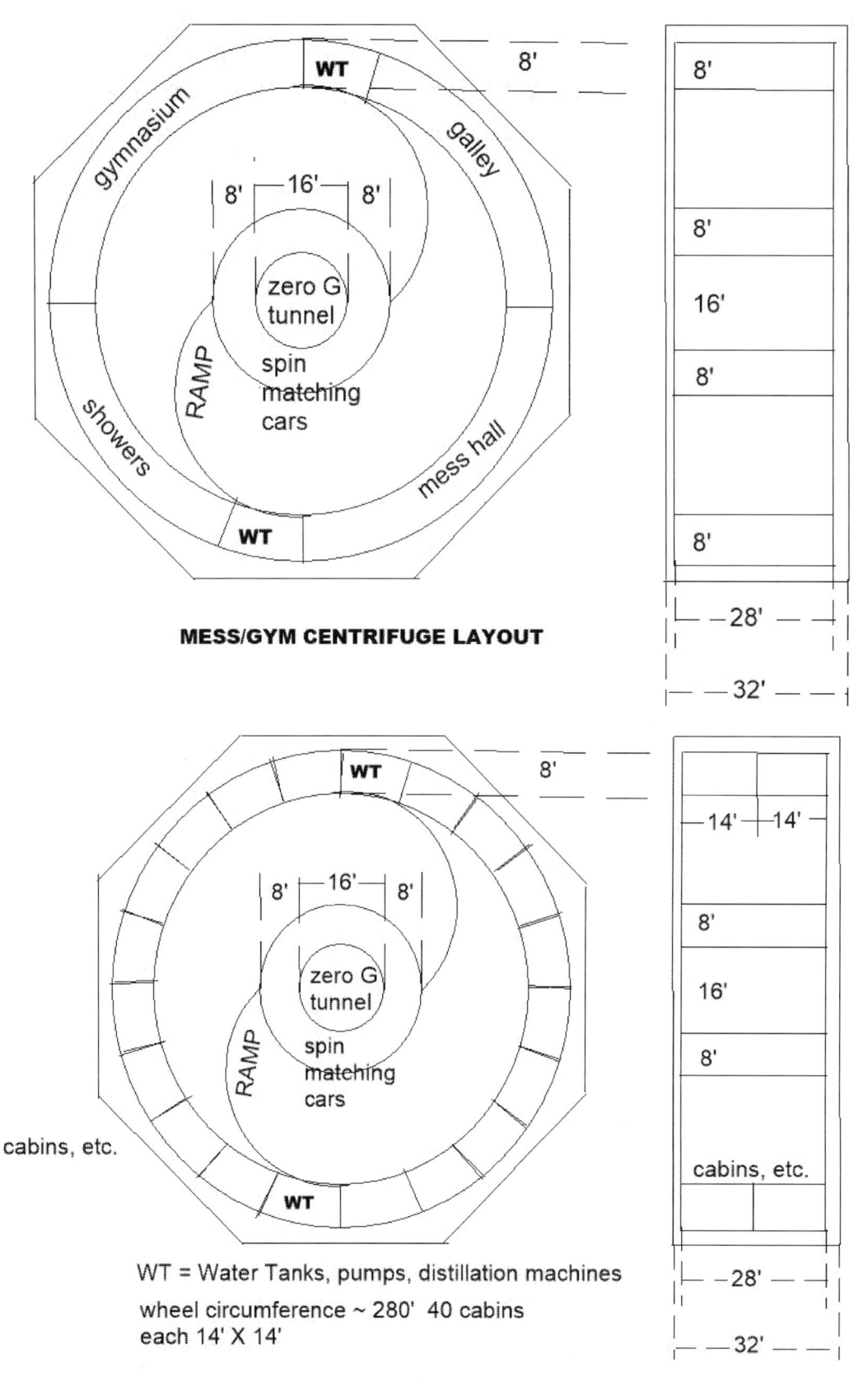

WT = Water Tanks, pumps, distillation machines
wheel circumference ~ 280' 40 cabins
each 14' X 14'

Fig. 16 Centrifuge cross sections

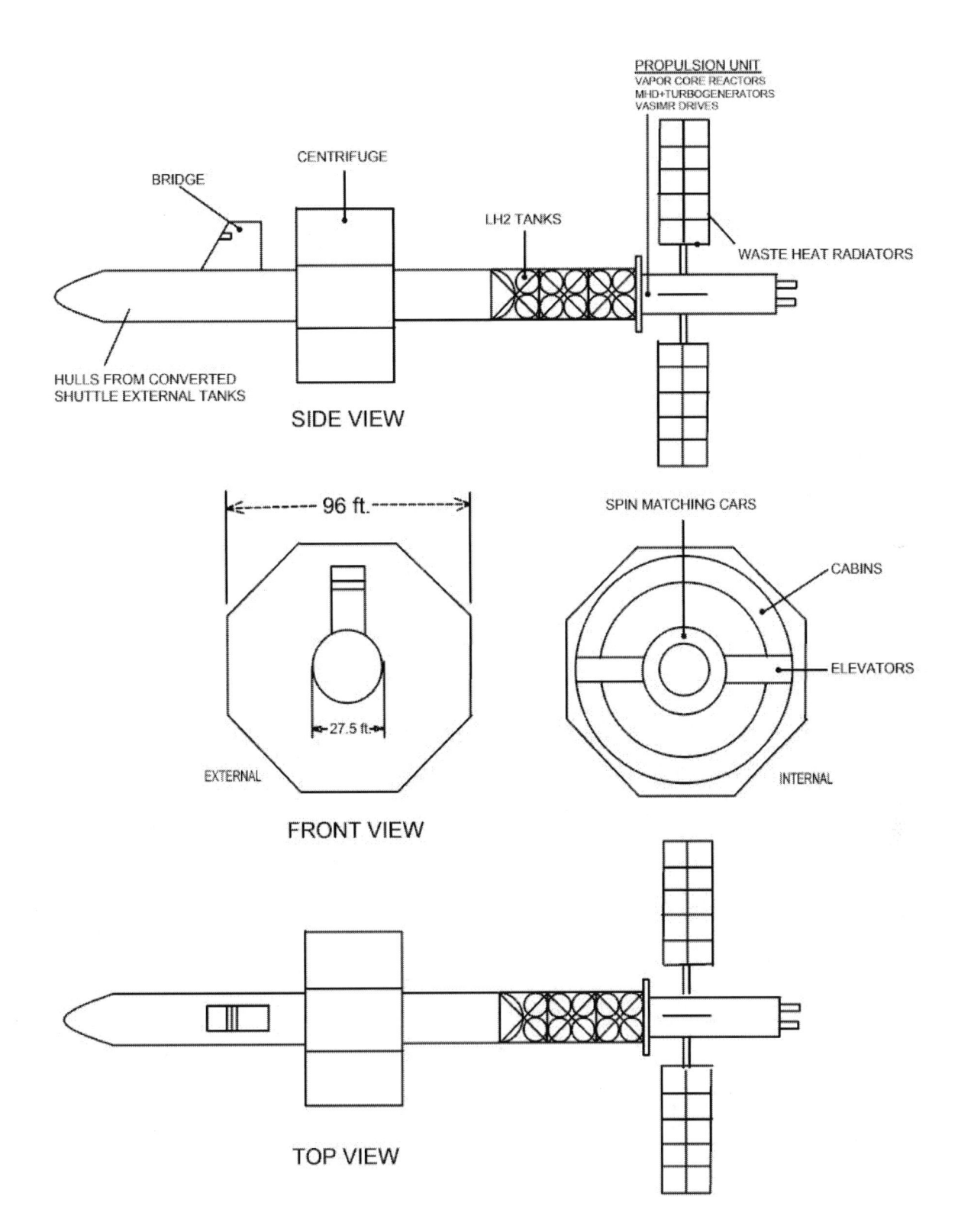

Fig. 17 Nuclear electric ship with one centrifuge countered by roll axis flywheels.

TWO COUNTER-ROTATING PASSENGER CENTRIFUGES

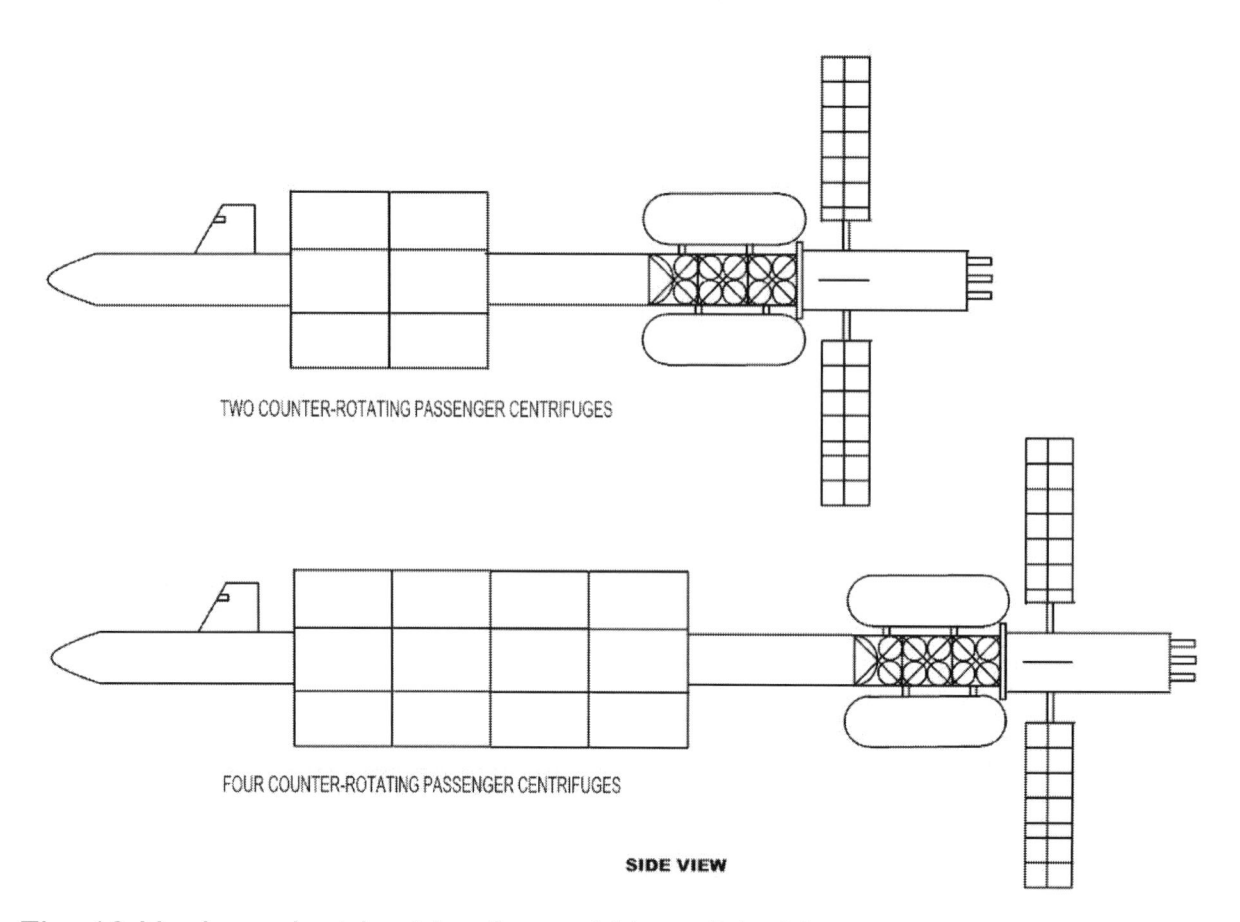

FOUR COUNTER-ROTATING PASSENGER CENTRIFUGES

SIDE VIEW

Fig. 18 Nuclear electric ships for rapid transit to Mars.

Interlunar ships could use chemical propulsion. Lunar liquid oxygen and silane mixed with metal powders could be used for propellants. This will not frighten those who are worried about nuclear reactors in orbit or nuclear ships retro-rocketing into LEO on their way back from the Moon. Governments will probably restrict space nuclear power. Hopefully, the law will permit nuclear ships in GEO. Chemical propulsion could get settlers to Mars in six months but nuclear electric propulsion could cut that down to 39 days. Marsbound travelers could ride space taxis from LEO to GEO where they board nuclear ships to Mars.

No hard evidence exists to back this claim, but 39 days in the lunar equivalent gravity of the centrifuges probably has no severe health consequences. As for travel beyond Mars, larger ships with larger centrifuges producing higher G force and active magnetic cosmic ray shields are called for. Someday, these ships will be built in space from materials mined in space. Their nuclear fuel will also be mined in space from the Moon and asteroids. They might use nuclear fission reactors and electric propulsion, or they might have helium 3 burning fusion drives. It could take many months even years to reach the outer solar system. A minimal amount of gravity will be needed to stay healthy on long space voyages. It might be necessary to produce 1 G and it might be possible to live at a reduced level of say 0.4 to 0.7 G. Nobody knows presently.

What will these ships cost? We can guess several hundred million dollars in today's money to several billion dollars. Success will depend on tapping space resources. It will also depend on the development of spaceship building techniques and technology in microgravity. Passage to Mars could use up a person's entire fortune. Settlers could sell their house, cars and all other possessions to book passage to Mars along with the needed equipment for living and prospering on Mars. Mars settlement would require some kind of organization and perhaps the sponsorship of wealthy individuals. There have been suggestions to dismantle the ship once it reaches Mars. In my view, expensive ships should be reused for as long as they will hold up and that could be many decades, even a century or more, in the rust and corrosion free environment of outer space. Lightweight nuclear ships would not be able to land on Mars. Settlers will descend from orbit in methane/LOX powered aerobraking capsules. Early shipments to Mars will consist of machines for tapping Mars' resources to make methane and LOX and reusable single staged vehicles that can rocket up to orbit and return to the surface by aerobraking, parachutes and retro-rockets.

Lunar tourism is foreseeable but Mars tourism would only be for the richest people, at least until a large industrial presence exists in space with a thriving economy based largely on automation and teleoperation; and a large number of humans living, working and becoming prosperous in space. Government backed efforts are not likely in free nations. Colonization is a thing of the past. Socialist experiments have no appeal. Groups of settlers might form corporations and take their earnings based on their share. Private homes and vehicles on the Moon and Mars seem reasonable. What to do with the unemployed or homeless? Find a way to make them earn their keep or ship them back to Earth!!!

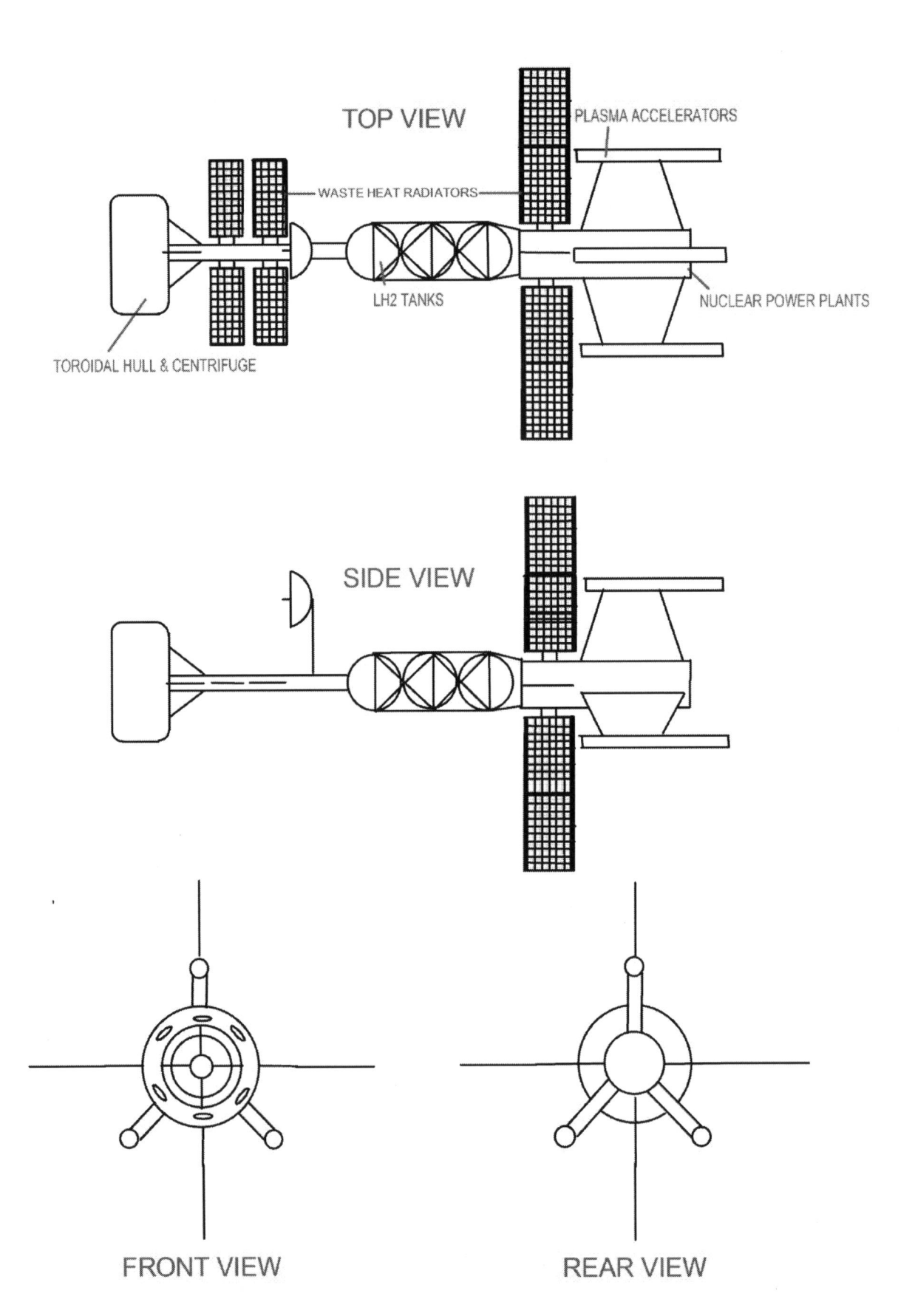

Fig. 19 Large space liner for travel to Mercury and the outer solar system.

In the previous chapter, I stated that about 350 tons of polyethylene shielding would be needed to protect passengers in the seating area from SPE radiations. Without going into the math, it will take about a thousand tons of polyethylene shielding for the two passenger centrifuges. The mess hall and gym centrifuges will have to be evacuated during solar storms in space. We can estimate that the ship will have a mass of about 1800 tons.

6 hulls at 30 tons each	180 tons
400 passengers	40 tons
Centrifuges	150 tons
Machinery	80 tons
Radiation shields	1,350 tons
TOTAL	1,800 tons

To reach the Moon in about 30 hours, a mass ratio of 4.48 was needed. That means this ship with centrifuges will need about 6,264 tons of silane and LOX. At a 1:2 mixture ratio, 2088 tons of silane and 4176 tons of LOX are called for. Tanks with a total of 1556 cubic meters for the fuel and about 3660 cubic meters for the oxidizer will be needed. Four tanks would each have about 1300 cubic meters of volume and an ET has a total of 2050 cubic meters and they are in the wrong proportions. The tanks would have to be made especially for this purpose and launched to LEO. About 30 tons per tank or "hull" is a good guess.

So now we are looking at $31,320 per passenger per round trip if we haul 400 passengers for propellant costs alone if the propellant only costs $1 per kilogram. This monster will never fly if there isn't lots of infrastructure in space and on the Moon. Ticket prices will be really high or propellant costs must come way down. Perhaps decades of development on the Moon with on-site materials rather than expensive imports from Earth and asteroid resources will make things cheaper.

Martians might have lighter ships. They could leave out the centrifuge shielding and just shield the forward hull. The forward hull could also be the storage area and they could get into a tubular space within the stores of water and food and use less polyethylene if there is a SPE. By using the forward hull as a storage area and solar storm shelter they could get the ship's mass down to just 800 tons. The four 30 ton silane and LOX tanks would be replaced by 120 tons of nuclear electric propulsion units and LH2 tanks.

Large Ships

In time, with extensive industrial development in space, entire spaceships will be built in space from lunar and asteroid materials. Perhaps some raw materials like copper and chlorine will come from Mars and complex, lightweight and electronic components will still come from Earth, unless semiconductor factories that take advantage of the vacuum are built in space. Large ships that can carry hundreds, maybe one or two thousand, passengers will be built in space for long voyages into the outer solar system. Some of these ships will carry settlers to Mars in greater numbers. Others will travel down to Mercury, a journey that will require a lot of energy due to the smallest planet's proximity to the Sun and high heliocentric orbital speed. Flights to Saturn, Pluto and the Kuiper Belt will take months, even years. Passengers will need "artificial gravity" in centrifuges and radiation shielding. A year in interplanetary space will expose travelers to 660 mSv of radiation from GCRs. A dose of one sievert will increase chances of developing a fatal cancer by 5.5%. Solar flare exposure is also more likely during these long voyages in space. Mass shielding consisting of tons of water or plastic per square meter of hull will be too heavy. Superconducting magnetic radiation shields that can deflect the high speed charged particles of cosmic rays and SPEs will be necessary.

What will these ships look like? Typically, ships with spherical hulls similar to the *Discovery* from the movie "2001: A Space Odyssey" are envisioned. There is a problem with ships like that. They don't offer the best configuration for magnetic radiation shields. It can be argued that the sphere is very strong and offers the best volume to surface area ratio, but that's not enough. One thing we can be certain of, these great ships will need nuclear propulsion with either fission or fusion to reach velocities high enough to reduce travel times and subsequent psychological stress from being confined to the ship, the ill effects of life in low gravity and radiation exposure. Shorter travel times will reduce life support requirements too. Water and oxygen will be recycled, but leakage is inevitable, so extra water and oxygen will be carried by spaceships. Ships will be too small for space farms. Food for the duration will have to be carried onboard. Naturally, shorter travel times mean less food must be carried.

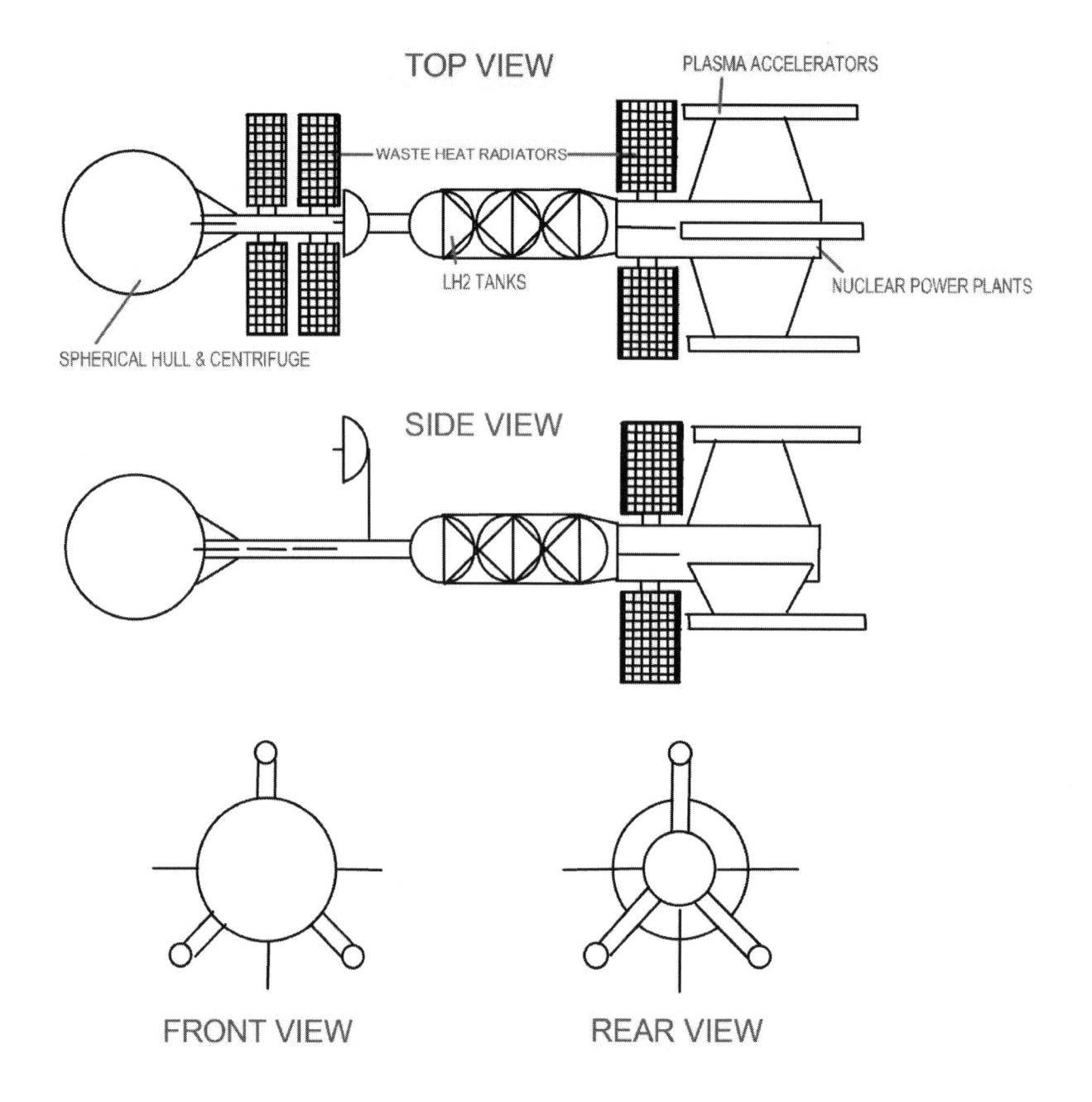

Fig. 20 Nuclear electric interplanetary vessel with spherical hull.

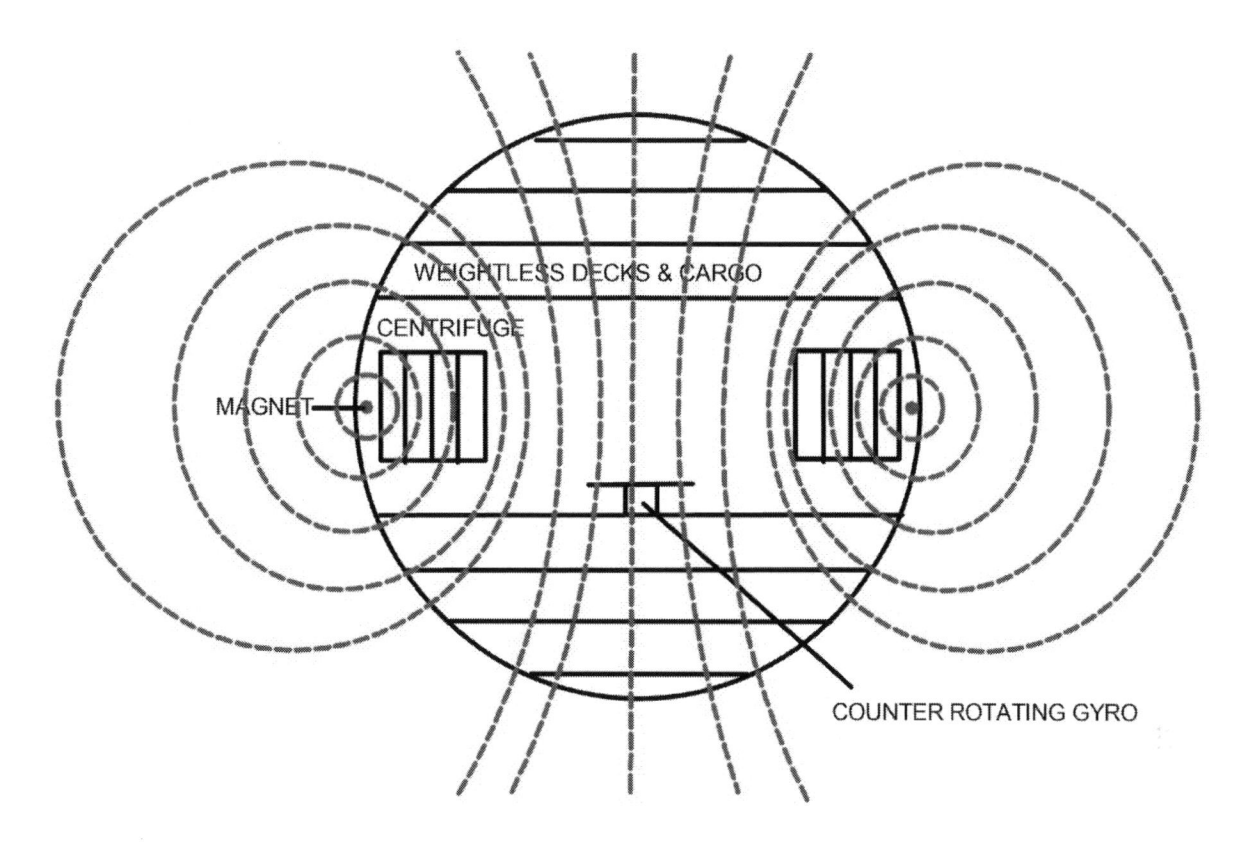

Fig. 21 Magnetic field surrounding spherical hull. Note the "polar zones" where charged particles can come barreling in and harm passengers.

A superconducting loop around the "equator" of the sphere could deflect charged particles from many directions, but not those entering in the "polar zones," for the same reason charged particles enter the polar regions of Earth's magnetic field and create auroras. The field lines almost perpendicular to the magnetic loop will allow particles to spiral along those lines and crash into the hull of the ship and emit secondary particle showers. There might be ways to solve this problem. Mass shields could be installed in the polar areas, but this will add a lot of weight to the ship. Stores of water and food could be placed in the polar areas, but how will they be effected by exposure to bombardment by GCRs over months and years?? Electrostatic shields might be installed in the polar areas but these might hog energy.

Fortunately, there is an entirely better geometry to work with—the torus. The torus allows us to have a magnetic shield with no habitable areas in the polar zones—just metallic structures. It allows us to put in a centrifuge that can be wider than one in a sphere so that we can get more Gee force at a lower RPM. The only problems I foresee, beyond the technical challenges of actually building such ships in space, are the biological effects of moving in a powerful magnetic field. This can cause subjects to experience dizziness, nausea, a metallic taste in the mouth and see flashes of light. It may be possible to surround the centrifuge with magnetic shielding. This is all speculation. There is almost no hard data on the ability of magnetic fields to deflect GCRs. There have been small scale experiments with particle accelerators. We don't know anything about prolonged living in low gravity and the biological effects of a powerful magnetic field even with magnetic shielding with mu metal, permalloy or Metglas. How much mass would magnetic shielding add to the ship? Torus and spherical hulls with centrifuges both share this problem.

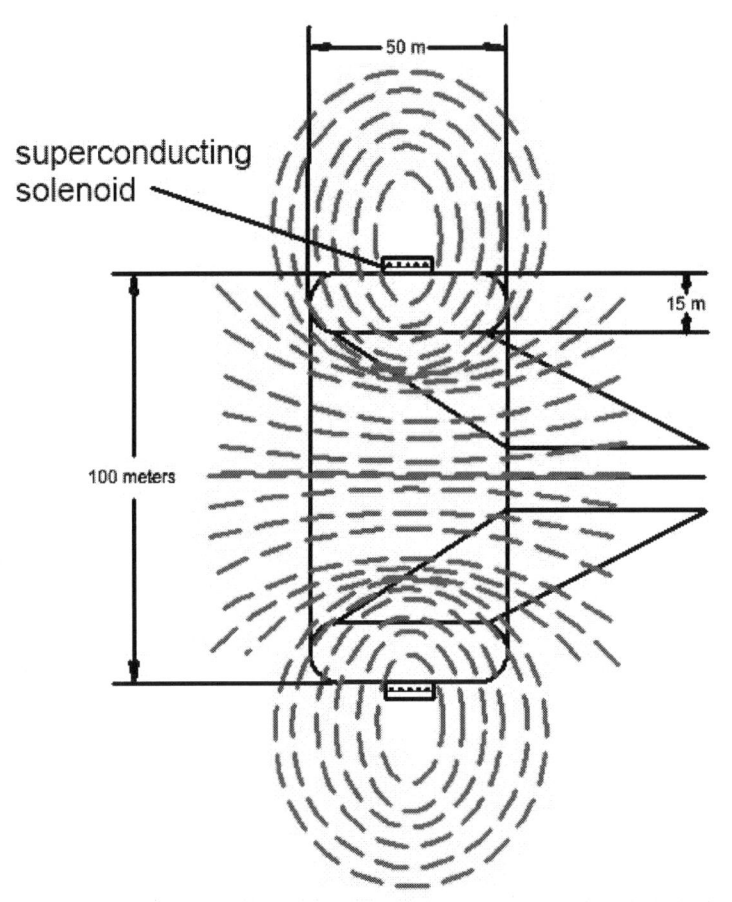

cross section of toroidal hull and magnetic field lines

Fig. 22 Toroidal hull geometry. There is nothing in the polar zones but metallic structure.

Top diagram labels:

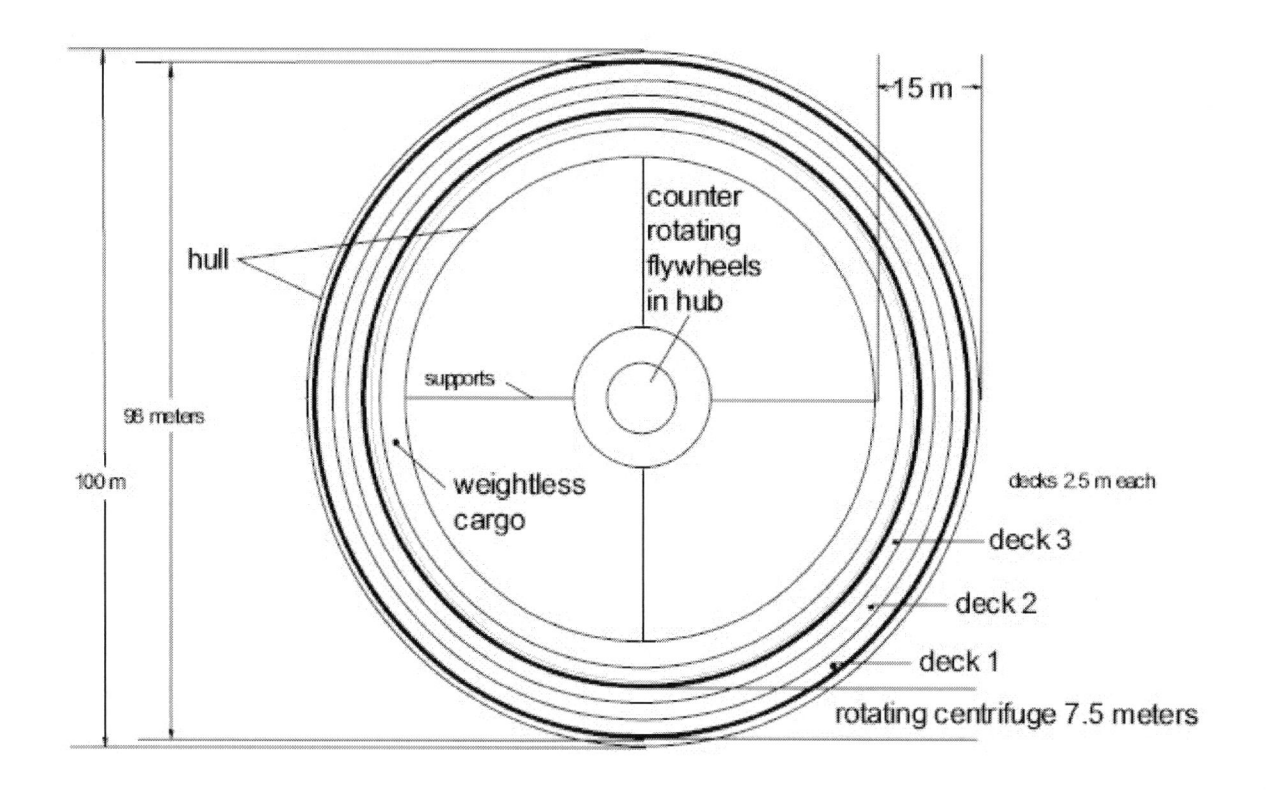

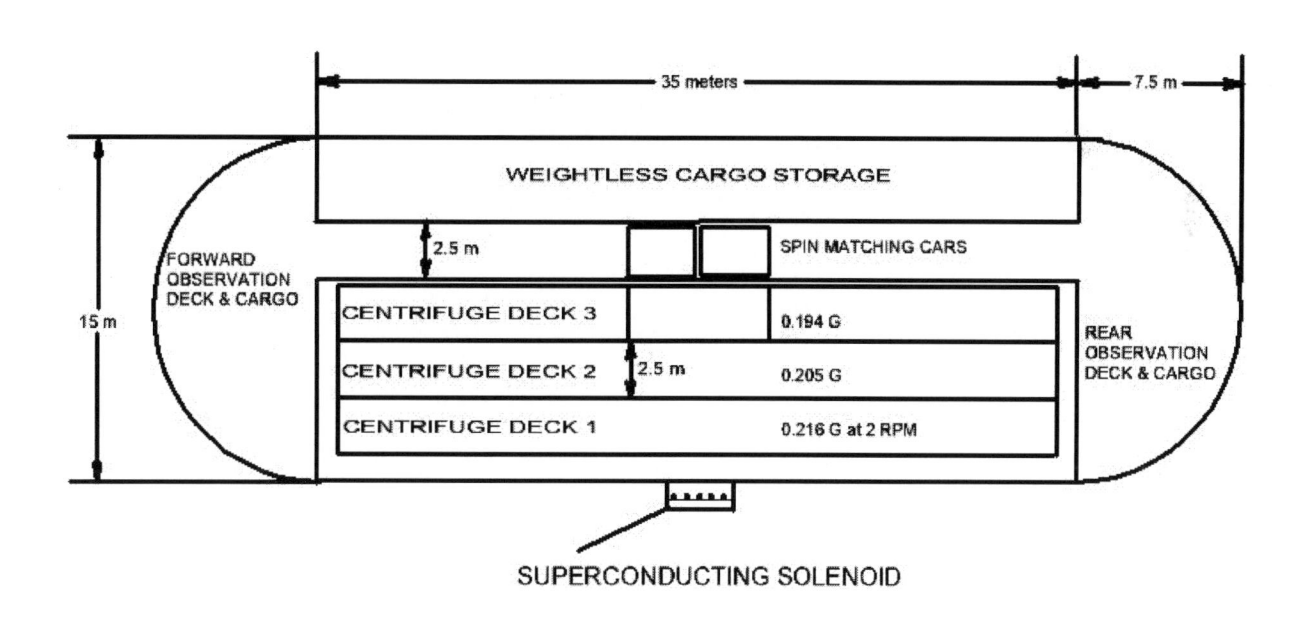

- hull
- 15 m
- counter rotating flywheels in hub
- supports
- 98 meters
- 100 m
- weightless cargo
- decks 2.5 m each
- deck 3
- deck 2
- deck 1
- rotating centrifuge 7.5 meters

Bottom diagram labels:

- 35 meters
- 7.5 m
- WEIGHTLESS CARGO STORAGE
- 2.5 m
- SPIN MATCHING CARS
- FORWARD OBSERVATION DECK & CARGO
- 15 m
- CENTRIFUGE DECK 3 — 0.194 G
- CENTRIFUGE DECK 2 — 0.205 G
- 2.5 m
- CENTRIFUGE DECK 1 — 0.216 G at 2 RPM
- REAR OBSERVATION DECK & CARGO
- SUPERCONDUCTING SOLENOID

CROSS SECTION OF CENTRIFUGE AND TORUS HULL

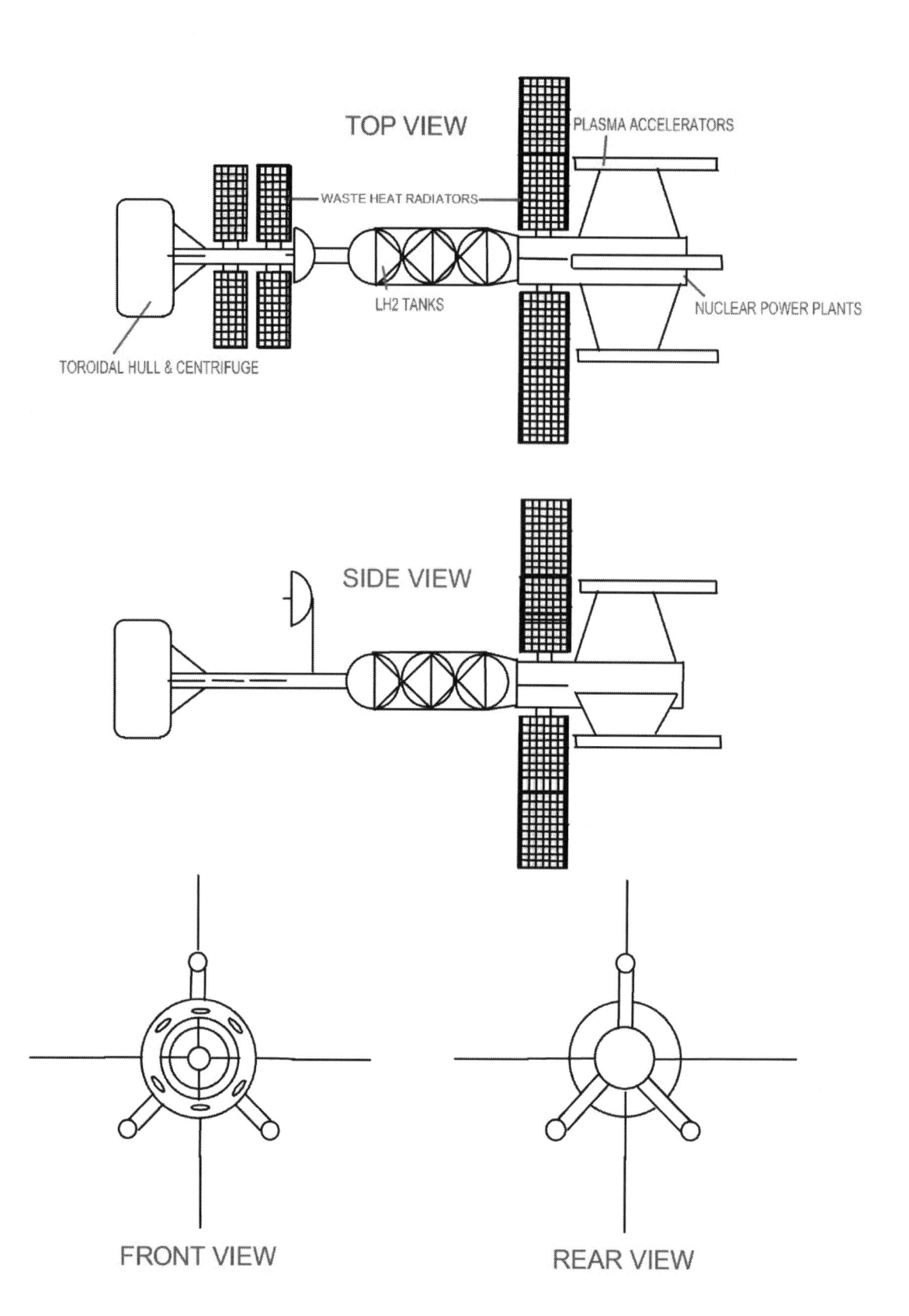

Fig. 23 Interplanetary ship based on torus shaped hull.

Supposing that the challenges of constructing magnetic radiation shields and centrifuges for large ships can be overcome, and that radiation levels can be reduced to acceptable levels for months' and years' long journeys, and humans can stay healthy in low G fields for equally long periods of time, perhaps with enough physical exercise, and even the use of medicines and nutrients, long term life support is still difficult. Spaceships that will be away for months and years will need life support technology that does not yet exist. Even the International Space Station needs periodic resupply.

Unless the ship is as big as a space settlement, there won't be enough room for crops, so food will have to be carried onboard. Most of that food will be dried and rehydrated with the same recycled water over and over again. An adult in a temperate climate needs 800 to 1600 milliliters of water, or an average of about 1000 ml, per day. An adult under thermal stress or doing a lot of physical exercise needs up to 6000 to 7000 ml per day.[11] Let's presume our crew consists of adults and that they exercise a lot to stay fit in the low Gee centrifuges available. They might need an average of about 2 liters (2000 ml) of water per day, just for drinking. Rehydrating food will demand even more water. If we have say a crew of 100 persons onboard a large ship traveling for six months to two years into the outer solar system, that would be 36,000 kg. to 146,000 kg. of water needed just for drinking. Then we need water for the stay at Titan or Pluto, and water for the return trip. We also need water for washing dishes and clothes, cooking/rehydrating food, mopping, showering, bathing, flushing toilets, etc. It seems that water requirements for a long voyage could add up to a thousand metric tons. Carrying a much smaller, lighter water supply and recycling it makes far more sense. Low flow shower heads, low flow flush toilets and gray water recycling systems should also be employed.

Many fruits and vegetables are over 90% water. Many milk and dairy products are 50% to 90% water. Nuts are only a few percent water. Many meats, fish and other seafoods are 40% to 70% water. Grains are only 17% to 23% water and just 10% to 12% after drying. Pasta, macaroni, spaghetti and noodles are 30% to 35% water and just 11% to 13% water after drying.[12] It seems it would be safe to say that food is on the average about 50% water. The average American, if there is such a thing, eats about 2000 pounds of food per year. If it was dried out that would be about 1000 pounds per year. So a ship with 100 people onboard will need about 50 english tons (100,000 pounds) of food per year. That's not nearly as much as the water requirement without recycling! Some of the food can be canned and frozen, since freeze dried and dehydrated foods are not always the tastiest. Perhaps with enough spices even the least appetizing freeze dried meats can be made more palatable. There are many foods available in the supermarket that we are accustomed to that take well to drying. Wheat flour, rye flour, corn meal, pasta, noodles, macaroni, spaghetti, raisins, powdered milk, powdered eggs for some people (many folks don't like them but I find them acceptable), beef jerky, Tang® and other foods will be just fine for making biscuits, muffins, cakes, bread, etc.

It could help psychologically if we have familiar foods to eat that are whipped up in a kitchen by good cooks instead of nothing but plastic and foil packaged space food bites. Besides spices, sugar or sweetener and salt will be needed. Dried fruit doesn't always have an appealing texture after rehydrating, but maybe it could be baked into pies, tarts and other goodies. We better bring some chocolate, coffee, tea and baker's yeast too. The most vital ingredient will be some real chefs and good cooks.

Oxygen is impossible to live without for more than a few minutes. The average human uses about 0.8 kg. a day. Our 100 passengers will use 80 kilograms per day or about 29,200 kg. per year. That's over 64,000 pounds! That doesn't seem to be as bad as the food and water supply problem, but if we go adding up the masses of water, food and oxygen for a journey to the outer solar system, a stay out there and the voyage home it becomes obvious that we must recycle everything we can. We can't recycle the food because there is not enough room for a space farm, although we could cultivate algae for part of our crew's nutrition, so we must recycle water and oxygen. The life support system could use algae to convert CO_2 back to oxygen or it could use a physio-chemical system based on a Sabatier reactor. Air could be passed through heat exchangers to freeze out water which are then defrosted and purged and another set of heat exchangers could get so cold they freeze the CO_2 out of the air. Those could be defrosted and purged and the CO_2 could then be pumped into the algae tanks or Sabatier reactors. Another way to isolate the CO_2 is to use membranes that pass oxygen but block the carbon dioxide. The advantage to using heat exchangers or membranes is that they won't become saturated like chemical adsorbents such as lithium hydroxide. There could be chemical adsorbents for emergencies and there could be reserve stores of liquid oxygen too. There better be a stock of spare parts for all parts of the life support systems, and technicians who know how to fix those systems.

SEMI-CLOSED SPACESHIP LIFE SUPPORT SYSTEM

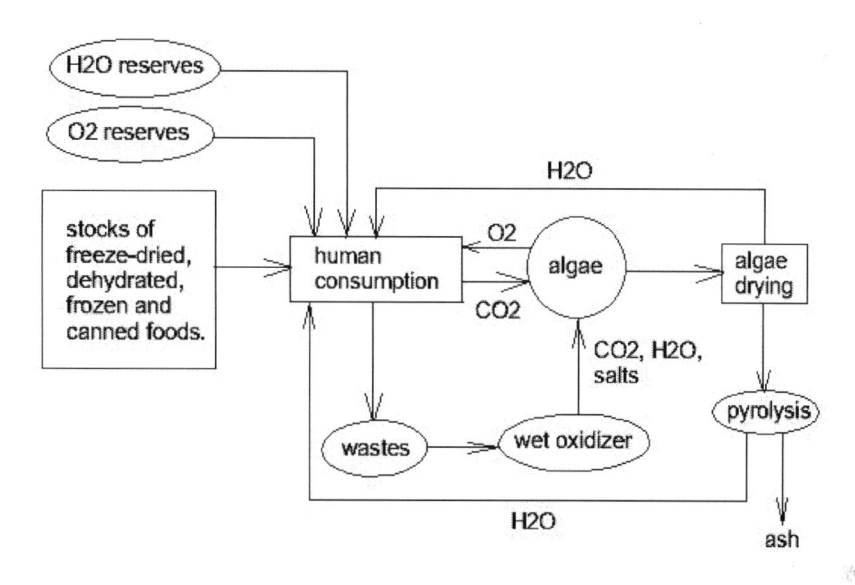

On a spaceship, all oxygen and water is recycled. Food becomes water and ash that is recycled in CELSS at large space stations and bases.

Wet oxidizer produces CO2, H2O, ammonium and mineral salts.
Pyrolysis recovers H2O content of algal biomass.

6 CO2 + 6 H2O --> C6H12O6 + 6 O2

C6H12O6 — pyrolysis --> 6 C + 6 H2O

This system does not recycle food; just oxygen and water.

Fig. 24 Bioregenerative life support system using algae.

SEMI-CLOSED SPACESHIP LIFE SUPPORT SYSTEM

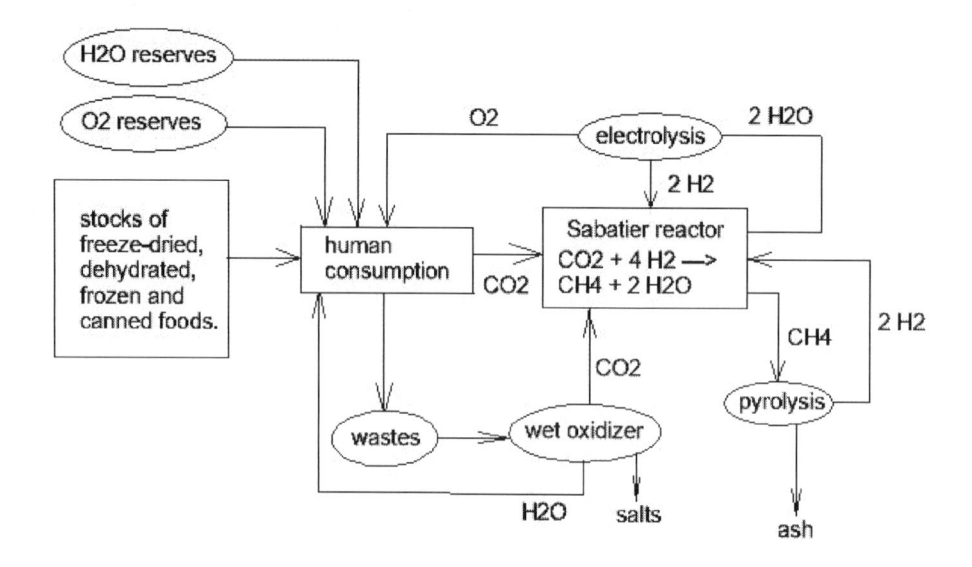

On a spaceship, all oxygen and water is recycled. Food becomes water and ash that is recycled in CELSS at large space stations and bases.

Wet oxidizer produces CO_2, H_2O, ammonium and mineral salts. Pyrolysis recovers hydrogen from methane. Electrolysis recovers oxygen from water formed in Sabatier reactor.

Wastes include exhaled H_2O. H_2O and CO_2 are extracted from cabin air with reversing heat exchangers or membranes.

This system does not recycle food; just oxygen and water.

Fig. 25 Physio-chemical life support system using Sabatier reactor.

Shorter voyages to the Moon or Mars are much less challenging. A ship with 200 persons on board that takes 30 hours to reach the Moon will only need about 500 kg. of drinking water, 620 kg. of whole, not dried, food and 200 kg. of oxygen. A ship with 200 persons on board that take 39 days to reach Mars only needs 15,600 kg. of drinking water, 19,430 kg. of whole food and 6,240 kg. of oxygen. That's much more than is required to reach the Moon on a tourist flight, but it's not all that bad. Still, it's enough to justify recycling of water and oxygen and storing at least part of the food in dried form. It's always possible that something could go wrong and the ship would be stranded in space far longer than 39 days before rescue. Recycling oxygen and water could make space travel much safer.

When we start talking about ships with 400 persons onboard or 1000 even 2000 passengers, recycling becomes 100% essential. Large space settlements with farms will be needed to supply these ships because shipping food, water and oxygen up from a planetary surface, even the Moon, would probably be very expensive. This helps makes the case for settlement of ELEO first with Kalpana type settlements just to provide food for lunar tourism.

Besides CO_2, there are other volatile organic compounds that get in the air of a spaceship or settlement. These could be frozen out of the air with heat exchangers or filtered out with membranes and catalytically decomposed to CO_2, water and nitrogen. They could also be cleansed out of the air with green plants. NASA has found that Areca plants, Snake plants, Money plants, Gerbera Daisy, Chinese Evergreen, Spider plants, Aloe Vera, Chrysanthemum and several other plants can remove all sorts of toxins from the air including ammonia, formaldehyde and benzene. [13] They are decorative too.

Power for Space Travel

The following discussion is based on approximations and educated guesses, but it will help us get "in the ballpark" when it comes to the enormous energies required for interplanetary travel. First, let us consider the kinetic energy equation:

K.E. = ½ mV² m is in kilograms V is meters per second and K.E. is in joules also called watt-seconds

Let's say we have a ship similar to that depicted in earlier chapters that has a mass of 800 metric tons and we want to reach a speed of 10 km./sec. for rapid flight to Mars.

½ (800,000 kg.) (10,000 m/s²) = 4 X 10^{13} joules or watt-seconds which when divided by 3600 sec./hr. is equal to 1.11 X 10^{10} watt hours or 1.11 X 10^7 kilowatt hours

That's a huge amount of energy.

A 200 MWe nuclear power plant would have to operate for 55.5 hours to produce that much energy.

This is only an approximation. The mass of the propellant would also have to be included, but since that changes during the acceleration integral calculus must be used, and we won't delve into that here. The purpose of this exercise is clear. Massive spaceships traveling at high velocities in the solar system will require very large amounts of energy.

What about a large ship with a mass of about 2000 tons and a delta V of about 30 km./sec. to reach Saturn in a year or two?

½ (2,000,000 kg.) (30,000 m/s²) = 9 x 10^{14} joules

Note that a one megaton nuclear bomb releases 4.12 X 10^{15} joules. This incredible energy equals 250,000,000 kilowatt hours. That's almost 25 times as much energy as our Mars ship needed. Our 200 MWe power plant would have to operate for 1250 hours or 52 days!!

To make matters worse, both ships have to carry more reaction mass along for braking into orbit around their destination planets. A faster trajectory means shorter transit times but it also means that the ship will approach the target at much higher speed upon arrival, so more high energy maneuvering is required. With an electric drive of some sort it will be necessary for the ship to brake slowly for weeks during its approach to the destination because of the low thrusts of electric drives and the huge amounts of energy that must be gradually delivered by the power plant.

The challenge doesn't stop there. If the ship and its crew ever want to come home, they must carry enough reaction mass for acceleration out of orbit from the target planet and they must brake upon return to Earth. The reaction mass, or propellant, could be equal to or greater than the mass of the ship. You can see why so many thinkers foresee small ships, slow speeds, long travel times and the employment of aerobraking instead of retro-rocketing at arrival. Many foresee one-way trips for the sake of colonization, or settlement. Settlement is a nicer word than colonization.

There are some ways around this. What if we don't carry enough reaction mass for return flight and get it from the destination? This is called ISRU (In-Situ Resource Utilization). Artificially intelligent robots could be sent ahead of time via minimum energy trajectories to extract resources and bootstrap facilities on the surface of Mars, Deimos or Phobos, or the moons of Saturn. A minimum energy trajectory or a Hohmann would take 260 days for Mars and six years for Saturn. The robots would have to be rather smart, but all they really have to do is make propellant. On Saturn's moon Titan they could just pump down atmosphere and pyrolyze it to obtain hydrogen for the electric drives. Hydrogen is most desirable because the lighter the particle the higher the exhaust velocity and monoatomic hydrogen makes the lightest atom. On Mars, the robots could dig up regolith and roast water out with waste heat from a small reactor. The water could then be put through electrolysis to get hydrogen. Simple tasks that will require thousands of lines of code!!!

What kind of power plants will be needed? An ultra-light vapor core reactor with MHD (Magneto-Hydro-Dynamic) power conversion has been designed by Travis Knight of Ne-Tech Inc., Samim Anghaie of the Innovative Nuclear Space Power Institute at the University of Florida, Blair Smith also of INSPI and Michael Houts of the Marshall Space Flight Center along with other researchers. They claim their power plant including the reactor, pumps,

MHD system, radiator and other parts for the complete system could produce more than one kilowatt per kilogram of mass. That would make the 200 MWe power plant weigh just 200 metric tons or less. Some designs included helium filled turbogenerators. With advances in technology, super-critical CO_2 filled turbines which are amazingly compact might be more effective.

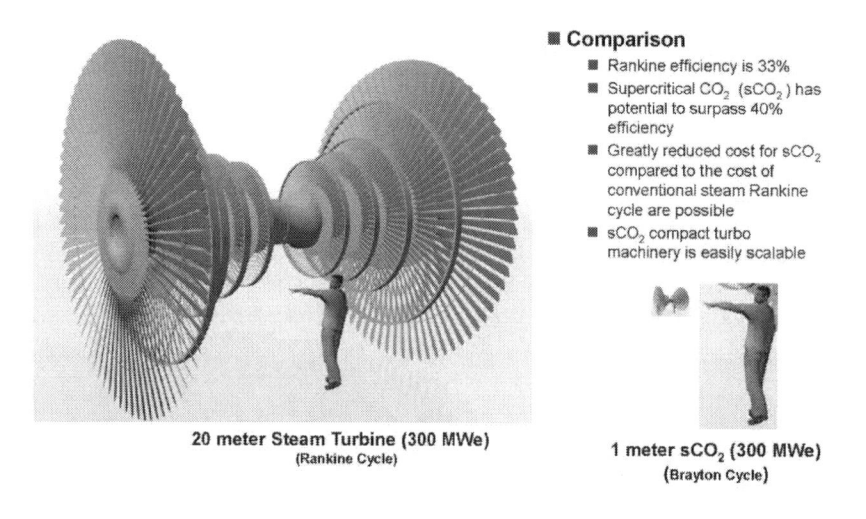

Fig. 26 Used by permission of Department of Energy

A bit of "handwavium" may be involved. At the temperatures the sCO_2 turbine must operate, 1500 to 3000 K., and pressures of several hundred bar, a significant fraction of the CO_2 will dissociate into carbon monoxide and oxygen. The turbine blades, housing and other parts might be made of tungsten, tantalum or rhenium coated with a ceramic. The problem is that oxide ceramics could be reduced by CO and carbides would be attacked by the hot O_2. The hot CO_2 could also react with carbides to form CO molecules. Perhaps borides or nitrides would fare better. There could also be exotic ceramics yet to be discovered by research into jet engines and hypersonic vehicles. If no material that can stand up to superhot CO_2, CO and O_2 can be found, a larger turbine using helium must be applied.

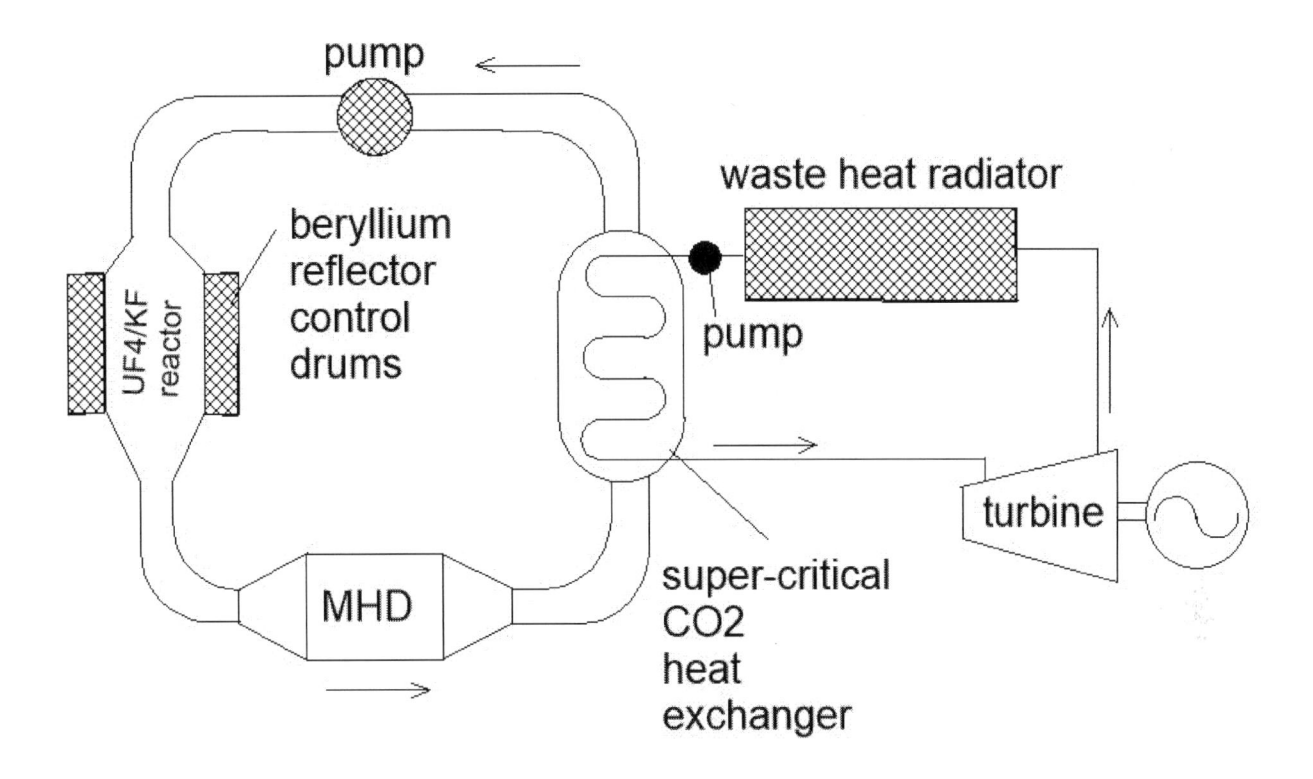

Fig. 27 Simplified diagram of Vapor Core Reactor/MHD/ sCO$_2$ turbine

Waste heat radiators must be thought about. The formula we need is:

P = e A 5.67E-8 T^4 e =emissivity   A = area T = absolute temperature

The temperature of the fissioning plasma in the reactor might be 4000 K. I don't know what this thing could be made of besides tungsten, tantalum, rhenium, carbon, thorium oxide and/or hafnium oxide. The reject heat temperature might be 1000 to 1500 C. If the system is 50% efficient and it generates 200 MWe, it will also need to reject 200 MWthermal.

200,000,000 = 0.95 A (5.67E-8) 1000^4 A = 3713 square meters

A square 61 meters (200 ft.) on a side.

As Robert Zubrin put it,"football field sized radiators."

Divided into 4 fins: 3713/4 = 928 m² 30m X 30m or 15m X 60m (50' X 200')

As for emissivity of 0.95 I am imagining our exotic tungsten or rhenium radiators with a coating of carbon particles impregnated by electro-thermal spraying.

If it rejects heat at 1500 C., then we get:

$200,000,000 = 0.95$ A $(5.67E-8)$ 1500^4 A = 733 sq. meters

A square 27 meters (89 ft.) on a side

Or four fins 22 ft. X 89 ft.

That's more like it.

The maximum Carnot efficiencies would be (4000 – 1500)/4000 = 62.5%

(4000 – 1000)/4000 = 75%

If these efficiencies could be reached then there would be less waste heat to reject and radiators could be smaller, however, lower reject temperatures mean less emission (T^4). Each fin will have two sides, so they should be half as large, but some heat will radiate away at an angle and contact other fins. Some sides of the fins will be exposed to the cold of outer space and other sides may face the Sun, so things will be more complicated. This is a very simple treatment of a complex subject. Hopefully, some young readers will be motivated to learn more about nuclear power, physics and mathematics for space travel.

That fissioning plasma at 4000 degrees Kelvin seems unbelievable. Reactor walls and other parts may need active cooling systems as well as construction with high melting temperature metals and ceramics.

Constructing Spaceships

Spaceship hulls or fuselages will be a lot like airplane fuselages. They will consist of longerons, struts, bulkheads, stringers, skins and other parts. There will be thousands of parts that must be assembled, riveted and welded together. It's also possible that entire hulls made of aluminum alloy will be printed at least partially with huge electron beam additive or selective laser sintering manufacturing devices in hangars filled with argon gas. The hulls will be fitted with seats, air ducts, life support machines, electrical systems, lithium batteries, plumbing and more by human and robotic workers on the ground, then rocketed into space.

In orbit, there will be platforms made of beams fastened together that have mechanical arms for grasping hulls and fitting them together. Teleoperated robots, orbital maneuvering vehicles and manned maneuvering units propelled by high pressure oxygen cold gas thrusters will also be employed. Most work will be done by teleoperated robots. Workers will only go EVA or "outside" in spacesuits when necessary.

External tanks will be held in place at similar platforms with large mechanical claws and arms with circular saws that have blades with black diamond teeth. They will cut the ETs into pieces. The curved pieces will be flattened out in rolling mills built largely in space from steel produced with lunar iron. The pieces will also be melted down, cast into billets and run through extruders to make aluminum rods that are then drawn through dies to make wire that is fed into 3D printers, perhaps electron beam additive manufacturing machines that work well in the free vacuum, to make parts of all kinds.

Centrifuges would be built in space. Flattened aluminum plates could be cut into desired shapes with CNC guided lasers, then held in jigs and fixtures and friction stir welded together. For centrifuges, some curved aluminum plates will be needed. Pieces from ETs could be rolled into the right shapes. It's also possible that large scale 3D printers will be used to make most of the centrifuge. That might be more efficient than having large numbers of space suited workers and teleoperators assembling thousands of parts for centrifuges. Large 3D printer device parts could be rocketed into orbit and assembled. They could have build envelopes with table areas of over 100 feet by 100 feet for printing up large parts of centrifuges like the octagonal outer bulkheads. These 3D printed one piece parts with reinforcing stringers and longerons will be very strong, simple and lightweight. There will have to be fixtures where centrifuge parts are held in place while they are welded together. There will have to be mechanical arms that can grasp and move the parts into place for the welding and other assembly tasks like riveting and bolting.

It might be possible to deploy huge plastic balloon hangars with internal pressure where ships are spray painted. It might also be possible to develop paints that won't just sublimate away in the vacuum. Electrothermal spraying of coatings like titanium dioxide powder on hulls might also be done. It might even be feasible to leave hulls uncoated and just polish the bare aluminum skins.

The interiors need work too. Cylindrical hull section interiors will be finished on the ground. They will be complete with lightweight velcro carpeting, painted and have other aesthetic touches. Cabins in centrifuges will be fitted with adhesive backed carpet tiles, plywood veneer on the plate aluminum walls and sound deadening ceiling tiles. Naturally much of this will have to come from Earth.

The reusable rocket engines, motorized gimbals, propellant tanks, pressurant tanks, piping and other parts will be assembled on the ground and rocketed into space in complete units that are attached to the rear hull of the ships. High efficiency multi-junction solar panels will be rocketed up, installed by workers in space and unfurled. Various antennas will also be attached.

Many parts can be made in space by 3D printing. Parts can also be made by casting and machining, by hand in pressurized workshops and with CNC machines. Casting would be done out in the vacuum while machining would be done inside where lubricants will not evaporate into the vacuum. Not only can parts for ships be made but also parts for machines, propellant depots, space stations and other things in space. With lunar materials including raw regolith, glass, basalt, cementious materials, silicon and metals along with titanium and aluminum alloys from ETs, and minimal amounts of plastics, chemicals and synthetic fibers as well as farm products just about anything can be made in space eventually. Ultimately, space manufacturing and construction will become completely independent of the Earth.

Light elements for plastics and such in limited amounts might come from lunar polar ices and solar wind implanted volatiles mining on the Moon in the early years. As space industry develops asteroid mining will become a big industry. Near Earth objects, asteroids and old comets, will supply Earth-Moon space with all the hydrogen, carbon and nitrogen that's needed. Even then it will be wise not to waste these resources. The use of silane and metal powders for rocket fuel to extend hydrogen supplies and the recycling of all plastics, composites, synthetic fibers and farm wastes will continue long after asteroid mining becomes a major space industry.

With carbon, hydrogen and nitrogen from asteroids, plastics and composites might become the major spaceship building materials instead of aluminum and titanium alloys. Anything that can be done to reduce the mass of a ship will be done. This will reduce propellant demands and cut costs. Lighter ships could also travel faster given the same amount of propellant used by a heavier, slower ship. If materials like carbon nanotubes and graphene ever live up to expectations it might be possible to make super strong ultra-light weight composites. Faster ships with shorter transit times to the Moon and Mars will please passengers, earn more money by making more flights over the course of their lifetimes and reduce radiation exposure times.

Bootstrapping for Spaceships

In my earlier book, Mining the Moon: Bootstrapping Space Industry, I discussed bootstrapping on the Moon and in space. Bootstrapping space stations for ship construction will be similar. The first spaceships will be built on the ground, rocketed into orbit, and assembled at stations consisting of frameworks with mechanical arms and an inflated section for human crews. The inflatable sections and solar panels will go up first, followed by cargo modules full of lightweight aluminum or composite beams that will be fastened together with bolts, rivets or by welding. Other parts, like claw sections, electric motors and more solar panels will come up later.

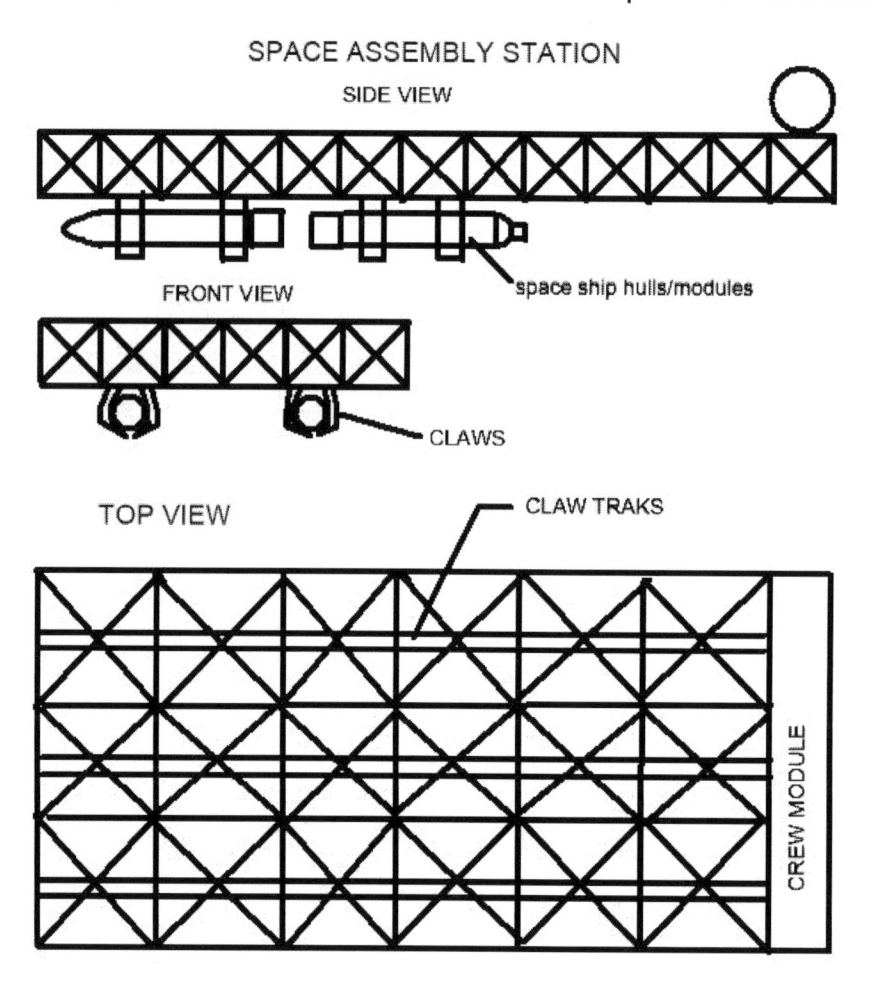

Fig. 28 Spaceship assembly station. Solar panels not shown.

Eventually, more assembly stations and ships with centrifuges will be desired. Stations that can produce centrifuges and more assembly stations could be built in orbit from lunar materials and old external tanks. At the same time, large space stations and settlements will be built in ELEO. Solar power satellite construction should be going on in GEO. Surplus oxygen, silicon, iron and calcium from SSPS construction activities can be hauled down from GEO with tugs using a combination of solar powered ion drives and propellantless electrodynamic tethers. There should be plenty of external tanks to supply aluminum and titanium alloys too. Someday, raw regolith could be hauled down to ELEO from the Moon and metals could be extracted from it for construction.

Iron bars can be packed in carbon power and roasted at red heat for several days. The carbon dissolves into the iron and forms steel. Steel would be too heavy for spaceships but it could be used for space stations. Silicon can be used to make solar panels that will endure less radiation exposure and subsequently less degradation in low orbit. Silicon and calcium by themselves don't have much structural use, but their oxides, glass and lime, do have construction uses. Large space stations could be built of steel, glass and concrete. New construction techniques for building in weightlessness will have to be developed. Concrete might seem like an unlikely material for space construction since it can't be cast in the vacuum. It could only be used inside pressurized structures. Concrete is rather heavy. Autoclaved aereated concrete, a very lightweight porous material made from lime, sand, water and aluminum powder, might be used preferentially.

Machines similar to large mass spectrometers could extract oxygen and metals from regolith including aluminum, magnesium and titanium for SSPS construction. In low orbit the main source of aluminum and titanium for spaceships and parts of stations will be external tanks. Tanks could be cut up with lasers or diamond saws and the pieces could be worked into other shapes or melted down and cast. The aluminum and titanium alloys from the tanks could also be powdered or drawn into wires and fed into 3D printers to make all sorts of things. The polyurethane insulation coated on the tanks could be removed and it could serve as source of carbon and hydrogen. Composite cargo modules could also serve as a carbon source.

Heavy equipment made of steel, like rolling mills and extruders, could be made in space instead of being rocketed up to LEO at great cost. There could be rotating stations made of frameworks of beams where sand molds could be made and heavy parts could be cast. These stations would only rotate fast enough to produce a fraction of one Gee; perhaps lunar equivalent. Rotation will provide "artificial gravity" that makes it easier to pour molten metals. In weightlessness some kind of injection molding system would be needed. The large parts, like rolls and extruder barrels, would be machined inside inflatables. Humans and robots would then assemble the parts. By making these things in space, hundreds if not thousands of tons of cargo will not have to be rocketed up from Earth. It will be possible to mass produce aluminum and magnesium plate, sheets, beams, pipes and rails in space once these machines are at work. Metal beams and plates could be welded up to create pressurized habitat and ship hulls. Rocketing beams into space to build space stations will no longer be necessary. Space industry will become largely independent from Earth. It will also become possible to make machines for mining asteroids in space instead of launching them from Earth.

Centrifuges for spaceships would be made in space from aluminum plate. They might also be printed. When there is enough infrastructure in space the toroidal hulls, centrifuges and other parts of much larger ships would be made in space. Materials would come from the Moon, near earth objects and perhaps from Mars. In time, there will even be factories that make electronic components and computers, textiles and garments including spacesuits. There won't be any wood and only small quantities of paper. Furniture will be made mostly of cast basalt, metal and some fiberglass.

A comparatively small amount of initial cargo, perhaps just a few hundred to a few thousand tons, using lunar and later other resources, will expand itself or bootstrap up to the large amount of infrastructure needed in space to build ships, stations, farms, construction facilities and more to support a growing space travel industry. Food and fuel will also be produced in space far more cheaply than by rocketing it up from Earth. A substantial population of workers will be sustained in space.

Spaceships and Health

Spaceships will contain large numbers of people in close confines. The spread of viruses and bacteria could endanger the health of passengers. It will be necessary to have air filtration systems that can remove 99.99% of germs and dust too. Smoking on board ships will be forbidden. The air filtration systems will have to have powerful fans so that all the air in the cabin can be filtered entirely several times an hour.

After each flight, the interiors will have to be steam cleaned and sterilized with UV light. Surfaces can be breeding grounds for disease as well as the air we breathe. Sewage tanks will have to be emptied and the waste treated in space stations, probably by wet oxidation which will require heavy equipment that generates a lot of heat. Wet oxidizer output will be sterile but could be processed further to get the water super clean psychologically as well as physically. The cleaning crews will have to wear haz-mat suits. They won't need a lot of education to do their jobs, but they will have to be good hard workers and be paid decent wages.

Spaceship crews and passengers must be screened for disease. Anybody with a cough or cold will have to have their trip rescheduled. They will have to provide proof that they have had all their vaccinations before flight. COVID-19 and variants of it may still be around in the future. Other diseases might be a problem for travelers from all over the world entering space in the future. Crews will be especially at risk, so they will need to be checked out by medical technicians before each flight. They will also need to shower up with antiseptic soap and wear fresh uniforms cleaned with bleach and perhaps UV treatment before each flight.

Space sickness could be a problem in weightless environments. Passengers might have gone to space camp and experienced Gee forces in centrifuges and weightlessness in airplanes. There may be tests to determine who is most susceptible to space sickness. Prophylactic medications may be necessary before launch. Crews will have to consist of people with a high resistance to space sickness.

Intense Gee forces will be experienced during launch and re-entry. People with heart conditions, high blood pressure, morbid obesity and poor physical condition will be discouraged or even forbidden to travel by rocket. A recent doctor's release may be necessary to buy a ticket. Diabetes might not be a show stopper, but severe mental illness and active addiction might keep some people from traveling in space. Alcohol might not be served and smoking anything of any kind will not be permitted. There might be bars in lunar resorts where drinking and smoking are permitted, but not in spaceships. Fire in space would be a disaster. It would be wise to make things out of fireproof materials like basalt fiber upholstery and cushion stuffing in addition to no smoking. Intoxication in space could not only lead to disturbing behavior but also to accidents. Could a drunk even use the zero-G vacuum toilet without mishap???

Pregnant and nursing mothers might be discouraged or forbidden from space flight too. Not only could Gee forces be difficult, but radiation can harm fetuses and there are limits to how heavily a ship can be shielded. The Russians launched a pregnant woman into space, supposedly, but she only went to low Earth orbit protected by the geomagnetic field. Infants might not take too well to weightlessness and changing their diapers would be very difficult. Even small children that have been toilet trained might present problems. They might get bored and start acting up. Perhaps there will be a minimum age for space travel beyond Earth orbit. Flight to LEO won't take long and staying in large orbital hotels or settlements won't confine kids too much. Going beyond Earth orbit might be too troublesome for kids under age ten and their parents. Mars settlers would want to take their children along, even small children, while others might prefer to wait for their kids to grow up some. Spaceliner companies will set their own rules, but governments won't be able to tell settlers with their own ships what they can and cannot do. Some people will be cautious while others will take stupid chances, get into trouble and become an example for others to avoid. If legal authorities tell people they can't do something they will do it anyway. When the fools get in trouble they will blame their leaders for not setting boundaries, so the people in power are damned if they do and damned if they don't.

Radiation will be a problem if there are solar flares. Ships will be shielded as discussed earlier. Travelers might consume radioprotective nutrients like vitamins A, C and E along with resveratrol, quercetin and curcumin for a few weeks before and during flight and a few weeks afterwards to prevent cancer. [14] More research needs to be done to be certain these nutrients will really work for passengers and especially for crews. Dosages must be determined. There might be cancer preventing prophylactic drugs and vaccines in the future as well as better cancer treatments. Cancer might be conquered someday as infectious diseases have today for the most part.

Pulsar Navigation

Over the years I have heard people suggest the use of pulsars as navigation beacons for ships in space and even ground vehicles on Venus. Well, NASA has already experimented with this. Experiments on the ISS with x-ray telescopes successfully located the space station's position.

Experiments on the ISS with the NICER (Neutron-star Interior Composition Explorer) x-ray telescope determined the station's position to within 3 miles. This experiment was called Station Explorer for X-ray Timing and Navigation Technology (SEXTANT). The NASA scientists predicted that they will eventually narrow that margin of error down to a few hundred feet.[15]

But wait a minute...isn't an x-ray telescope for tracking pulsars a massive piece of equipment?? The mirror on the Chandra X-ray telescope launched in 1999 had a mass of 18.5 metric tons per square meter.[16] The instrument used during the SEXTANT experiment only had a mass of 5 kilograms.[17] There has been a lot of progress in the field of X-ray optics. Pulsars emit in the radio portion of the spectrum, but a radio telescope attached to a spaceship would be rather large. Very few pulsars emit in the optical region (only five optical pulsars are known) and an impractically large optical telescope would be needed.[18] Compact X-ray telescopes win.

Presently, space probes are tracked and navigated by the DSN (Deep Space Network) using large dish antennas and radio. This won't suffice in the future when hundreds, even thousands, of ships are out in the solar system. To make things worse, the accuracy of radio tracking and navigation decreases with distance. Communication relay stations will need to know their location accurately so that lasers can be aimed precisely at them and at distant ships, colonies or worlds. Using pulsars for a sort of "solar system GPS" seems to be the wave of the future. The future communication laser beams will probably contain data pertaining to the location, trajectory and velocity of the ships and stations as well as text messages, email, faxes, voice and video messages and even internet searches and streaming content. Even in space we will want to stay connected and we definitely don't want our spaceships to get lost!

Collision Avoidance in Space

The chances of colliding with another spacecraft, satellite or old upper stage in planetary orbit are low. The chances of colliding with an asteroid or being hit by a meteor in interplanetary space are also low. However, they are not non-existent. Nobody was worried very much about the chances of the "un-sinkable" ship *Titanic* crashing into an iceberg. They were so lax that the builders of the great ship didn't include enough lifeboats. Only fools would not try to avoid a repeat of this disaster in space. Someday, there will be spaceships carrying hundreds even thousands of passengers and valuable cargoes including precious possessions of the wealthy. It would be awful if a luxury space liner had to collide with a space rock and suffer hundreds, even thousands, of deaths just to wake people up to this danger of space travel and get them to do something about it.

Radar systems that can scan a spherical volume of space with a radius of several thousand miles surrounding the ship are called for. Synthetic aperture radar (SAR) might do the job. As of 2010, airborne systems provide resolutions of about 10 cm, ultra-wideband systems provide resolutions of a few millimeters, and experimental terahertz SAR has provided sub-millimeter resolution in the laboratory. [19] Detecting and avoiding even small rocks will be necessary. What if a small meteor broke a window and rapid depressurization resulted? What if a somewhat larger meteor crashed into the ship's nuclear reactor and crippled the propulsion system?

Ships could travel in convoys with enough extra space onboard to take on passengers from a crippled ship. Since launch windows to and from other worlds are often brief in time and months if not years apart, it would make sense for convoys to depart at roughly the same time. Lifeboats won't help much. I seriously doubt that small life boats could carry enough oxygen and other supplies needed to sustain people for weeks or months in space. Some kind of inflatable "escape pods" might be helpful for passengers abandoning ship and moving to another ship in the convoy. That should only take a few hours. Even so, meteor impacts and collisions with asteroids or old satellites should be avoided.

Synthetic aperture radars could scan for objects of all sizes and determine their size and potential danger. The radars would have phased array antennas for rapid electronic scanning instead of mechanical scanning with dish antennas. They might also use Doppler data to determine the object's velocity along with other radar data to determine distance, position and trajectory of the meteor or piece of space junk and feed that data into computers. If the computers determine that a collision is imminent they could fire thrusters or orbital maneuvering system rockets to get the ship out of the way without excessively altering the ship's course in a way that would cause it to miss its destination by thousands of miles or less. The computers would also alert the crew but they wouldn't wait for slow humans to redirect the ship. There could also be radar buoys in space that scan for dangerous objects and transmit data via relay stations with radio or laser to ships in space. A central data bank containing information about the orbits of just about every space rock down to less than 1 cm in diameter and pieces of space junk could be created and updated constantly. This data would be used when calculating a ship's course through space to ward off collisions.

When tens of thousands of ships are plying the space of the solar system knowing their course or trajectory ahead of time when planning a ship's course could also be a lifesaver. Collisions of large ships traveling at high speed, manned or robotic, would probably be fatal to all onboard. The cost of losing a robotic ship with its freight would also hurt someone's bottom line, and that's not desirable. Ships that have no flight plan stored in the central computer data bank will be highly suspicious. Radar buoys might locate them and alert the authorities. Ships would probably be fitted with transponders too. Ships detected by radar with their transponders turned off would also be suspect. The collision avoidance systems could also help prevent smuggling and human trafficking in the future.

The computer power would be immense and the code would be millions of lines long. The mechanical and electronic components of the radar dishes and transceivers would also be impressive. Whoever defeats all these challenges and manufactures a product that could be required by law to be fitted on every spaceship that carries humans could make a lot of money in a future time when millions of people travel in space every year, or every day! Robotic cargo ships might not be required by law to have these systems but the insurance companies might require it. The law might require robotic ships carrying emergency supplies of food or medicines to have collision avoidance radars also.

In addition to radars, ships could be divided into airtight compartments linked by airlocks. If by some chance a meteor did breach the hull some of the passengers would be safe. Hulls might have layers of self-sealing materials to reduce the damage done by micro-meteors. Inflatable "life boats" or life pods could be kept on board for use in an emergency that allow passengers and crews to escape to another ship in the convoy. Traveling in convoys of less than three ships will probably not be recommended.

Every effort must be made to make space travel safe, and that includes increasing launch rocket reliability along with escape rocket systems, using two-stage landers with back-up rocket motors, and improving re-entry and landing systems. Space travelers and shippers will definitely want insurance policies. Sick bays will need equipment for handling decompression sickness due to accidents that involve loss of pressure in ships and in spacesuits. Space travelers must be prepared for every mishap, since Space Fleet ships will be unable to race to the rescue at warp speed in the real universe.

Interlunar Rocket Safety Mechanisms

If we are going to have lunar tourism we must have safety. Rockets must operate reliably, especially when retro-rocketing into lunar or Earth orbit. Getting to the point of this document, it will be better to have fuel valves that fail shut rather than open so that we don't lose control of the rockets. If a rocket motor casing burns thru or a valve fails shut and deactivates the motor we can have fuel lines with valves and electric pumps that connect the propellant tanks so that fuel and LOX from the failed motor/tank module can be pumped to other tanks that still have working rocket motors. With 4 motors we are safe, with 3 things are acceptable, with 2 things are dangerous and with just 1 things are impossible. If one motor fails we will have to fire other motors longer to perform maneuver. Then when the thing docks it has to be repaired.

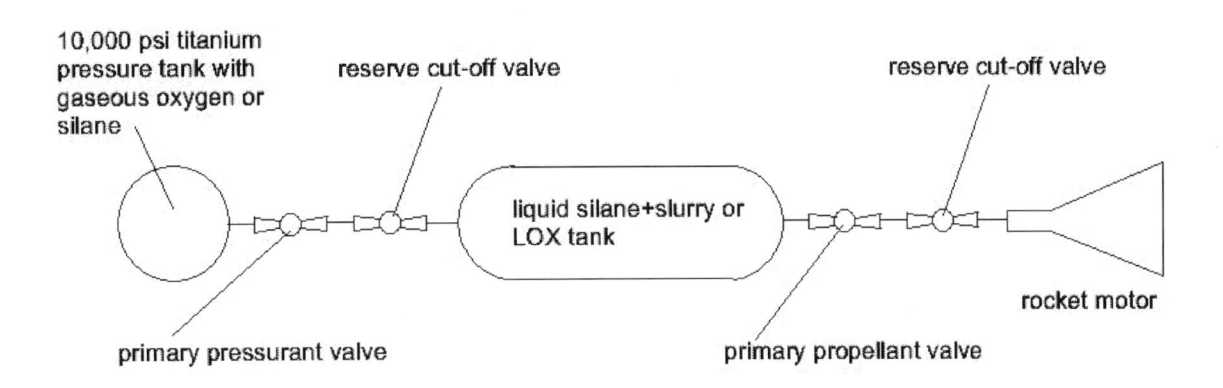

Fig. 29 Reserve cut-off valves for safety.

The rockets will use simple pressure feed. Silane (SiH_4) and oxygen will form water and silica when they burn. With metal powder suspended in the silane, particles of metal oxides will form too. With a turbine to drive pumps and hydraulic steering systems, there could be problems with silica and metal oxide deposits forming on the turbine blades and ruining them, thereby necessitating costly service and repairs. So pressure feed is preferrable with this propellant combination of silane/metals/LOX, but centrifugal pumps with electric motor drives might also be used if the ship can carry enough batteries for power.

The rocket motor would be steered with electric motorized jack screws that move it around on gimbals to steer. If something went crazy we'd want to shut that motor down rather than let it steer us the wrong way or push the ship into the wrong attitude. The other three motors would then take over. It's a rather simple system so it should be reliable and cost effective to service and maintain.

The valves will be electric motor or solenoid controlled and spring loaded so that if the power to the motor is lost the valve gets stuck closed. The valve motor or solenoid will be energized by current from a relay that is controlled by computers. If due to a computer glitch, or the relay gets stuck or there is a short circuit and the power doesn't shut off the valve could get stuck open. So electrical malfunctions as well as mechanical sticking are sources of concern.

Since we'd rather have the motor fail off instead of on and send us out of control we have reserve cut-off valves in case the primary valves get stuck open. If the valves get stuck closed we have to pump fuel and oxidizer to other tanks to perform maneuvers with rocket motors that are still working. **The valves could and should be spring loaded so that power loss to the motor or solenoid that opens and closes and perhaps throttles flow thru the valve will cause the valve to close. It is also possible that the power to the motor or solenoid won't turn off and the valve won't close, so there should be a kill switch or circuit breaker in addition to the reserve cut-off valve.**

I want to use ball valves with Teflon or PFA coatings so they are slick and less likely to stick. Teflon works at cryogenic temps. We might even have drilled passages and inject oil into the ball and body surfaces for extra slipperiness and lower sticking possibility, if necessary.

I want pressurant reserve cut-off valves so if the primary pressurant valve sticks open we can shut off the gas and not over pressurize the propellant tank. We might need some kind of regulator in there also.

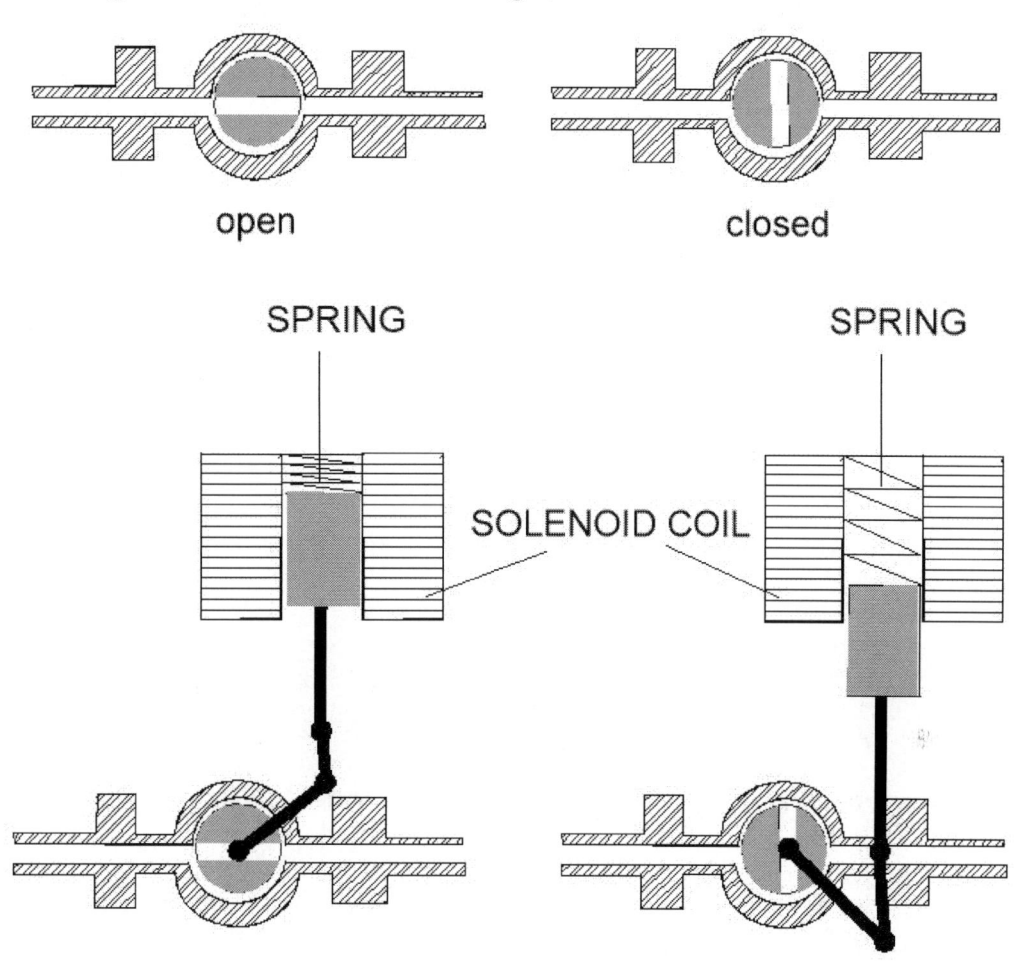

Fig. 30 Ball valves. The seals would be Teflon or PFA...and there could be an oiling system to prevent sticking.

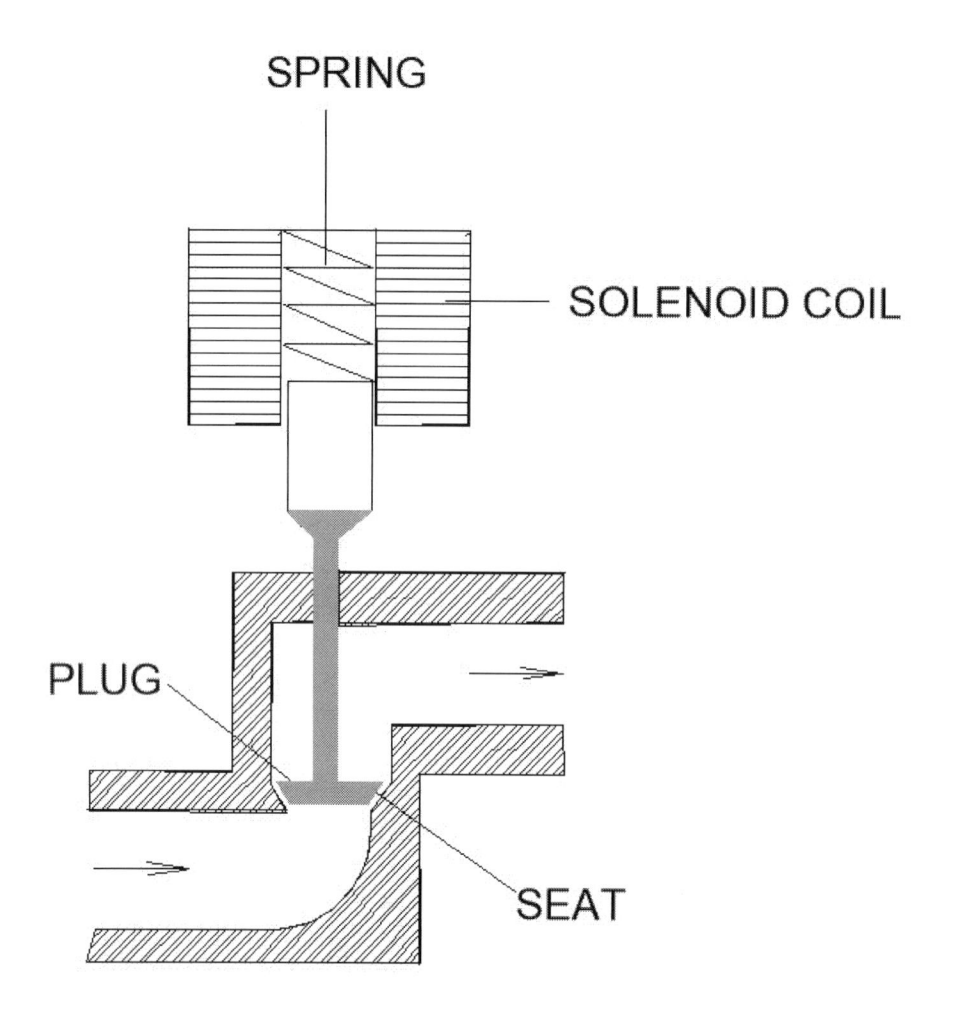

Fig. 31

This kind of valve is not wanted. The pressurized fluid from the left could push the plug off its seat and get stuck open. To make matters worse, the fuel is a mixture of silane and metal powder, a slurry, and the metal particles could get in between the plug and the valve seat and not seal….then we'd have fuel leaking through especially if it was under pressure. LOX won't have this problem, but I'd still rather have the LOX valve fail closed instead of open so all the LOX doesn't get spewed away into space!!!!

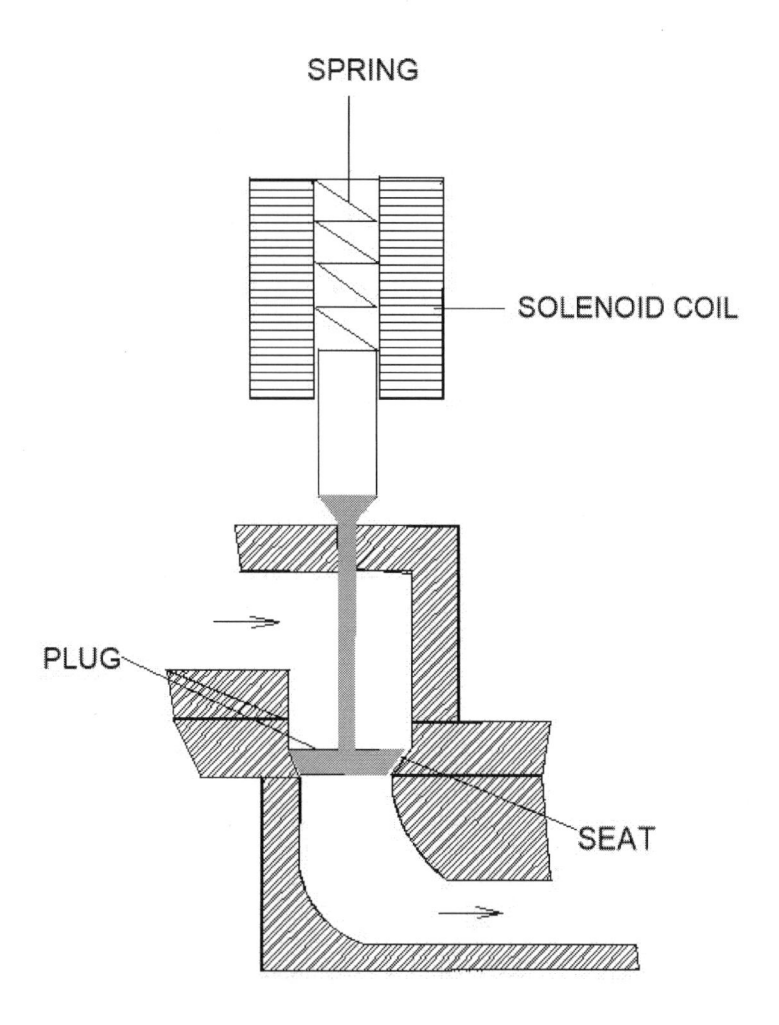

Fig. 32

This valve can be safer. If the solenoid fails due to power loss the spring and the pressure forces the valve closed, so we don't lose control. If for some reason the current doesn't turn off when we want to close the valve we have a problem, so we need a reserve cut-off valve as pictured above so the ship doesn't go crazy or propellant isn't dumped into space. Thrusters will need similar arrangements because we don't want stuck open thrusters as well as main rocket motors. But even this valve is not desirable because the metal particles from slurry fuel will get stuck between plug and valve seat and fail to seal. With thrusters and ullage motors running gaseous silane and oxygen this won't be a problem, but we'd still want an arrangement where spring and pressure tends to shut the valve closed.

So all these backups and safety features mean a more complex spaceship with more to go wrong....Murphy's law is real and going into space may be as dangerous as going into combat so you don't want to leave any stones unturned. Engineers will need to design and manufacture to close tolerances and test things for like hundreds of thousands of trials to make sure the stuff works for a long time without mishap.

When we leave orbit or retro into orbit we will have to fire the ullage motors to settle the propellant in the tanks. There will have to be some kind of agitator in the slurry fuel tank to keep the metal powder all mixed up in the silane. The agitator could be rotated magnetically instead of having a penetration thru the aluminum tank walls. And we'd need a solar shield so liquid doesn't contact hot tank walls and boil up. Then we have to open reserve pressurant valve and primary pressurant valve to push the propellant out and then open the reserve propellant valve and then at the right instant open the primary propellant valves to the motor and fire the engines. Each motor has two tanks, one for fuel and one for lox, and each tank has 4 valves, so each motor relies on 8 valves and four motor/tank modules means 32 valves! And they all have to be controlled electrically by a computer that gets everything open and closed in the right sequence to make rocket motors fire and turn off at precise moments within fractions of a second......I guess there would have to be flow sensors in the pipes so the computer would know when a valve is stuck open or closed.

Silane is pyrophoric, meaning it ignites when it contacts oxygen in the air. Gaseous silane and oxygen to the thrusters and ullage motors will probably ignite without an electric spark, so these will be very reliable. Will supercold liquid oxygen and liquid silane ignite upon contact or is an electric spark required??? If no spark is needed, rocket motor ignition will be very reliable.

Spaceship Accommodations

A typical airliner seat is less than two feet wide. For the ships described previously I imagined seats three feet wide and five feet deep. With smaller seats it would be possible to cram more passengers into the ship and make more money. It could also be possible to reduce ticket prices. However, I wanted to make the seats wide and comfortable for 30 hours including sleeping. That's a long time to be strapped into a seat even with occasional limb stretching in the floating rooms. In weightlessness passengers should not become sore sitting so long. Businessmen of the future might shrink the seats, add more passengers and even replace the floating rooms with seats just to get more people on board.

A system of antennas connected by coaxial cable wired up to the ship's telecommunication hardware will allow cellphone use throughout the ship. Passengers might have to download an app and pay for the service as they do on cruise ships today. It will be possible to send and receive voice mails, video, text and email. Internet surfing might be possible for ships traveling between the Earth and Moon if passengers can tolerate the 3 second lag time. In centrifuges the aluminum plate walls, floors and ceilings will block radio signals. An antenna will have to be located in every cabin so that passengers can use their phones. There will also be antennas in the centrifuge dining halls, game rooms, etc.

Cabins in centrifuges will have dimensions of about 14 feet by 14 feet. There will be a flush toilet closet in each cabin. Air will be thoroughly filtered and sterilized. Cabins and common areas will have decorative plants illuminated by LEDs that also cleanse the air. Metal walls will be covered by varnished plywood veneer. Floors will be carpeted and acoustic tiles will cover the ceiling. Walls between cabins will consist of double metal sheets or thin plates stuffed with basalt fiber sound deadener for privacy. There will be two couches with fold out queen sized beds for a family of four, two couples or a party of four. Some cabins might also have fold out bunk beds for larger groups. Each cabin will have a big flat screen TV with access to thousands of movies in the ship's computers. It's also possible for passengers to have handheld devices with solid state hard drives that contain thousands of their favorite movies, songs and books that can be hooked up to the TV. There will also be a microwave oven, coffee maker and a mini-fridge. Room service might be available too. Larger ships of the more distant future might have a conveyor belt and dumbwaiter system in the centrifuges to deliver meals, drinks and return dinnerware. It might not be a food synthesizer, but it will make it possible to get meals prepared by top chefs without leaving one's cabin.

It might be possible to put a shower stall in cabins. I have imagined a shower area next to the work out gyms with private shower stalls and changing rooms. The water would be recycled. It would be filtered, distilled and sterilized. After each flight the shower water tanks would be drained and replaced with water from the space dock station. The water might be clean enough for several flights, but it will be more psychologically appealing if it is changed more often. The water in the flush toilets will also be recycled and the sewage tanks emptied after each flight. The space dock stations will be very large and contain space farms. Water transpired by the crops will be condensed from the air and purified before use for drinking, cooking, showering or flushing toilets.

After each flight maids will change all the linen and wash it in very hot water, possibly with bleach. It's also possible that UV light will be used to sanitize the linen. The ships' interiors will be steam cleaned and sanitized with UV light and disinfectants like hydrogen peroxide. For long flights to Mars, travelers will have to keep their cabins clean themselves and go to the laundry to do their own sheets and clothing. It won't be economical to put a staff of maids and custodians on each ship bound for Mars or more distant locations in the solar system. Recycled water will stay clean enough for a 39 day voyage to Mars and travelers to Mars will be a hardier breed who don't mind recycled water. That's all they are going to have on Mars for many years!

Interlunar travelers will be offered weightless seats, steerage with up to eight passengers per cabin, standard cabins with four people and first class cabins for two. Mars bound travelers might enjoy a limit of two people or a family of four per cabin given the longer period of time they will be in space. Crews will be on duty for eight or twelve hours per day. They will work day/evening/night or day/night shifts. It will be possible for them to share bunks. Crews will be less concerned with luxury and more concerned with basic comfort and all the money they are making. Nobody is going to do the job if they aren't well compensated. People beyond their childbearing years with very little family history of cancer will be preferred for crews. They will have to be multi-talented. Individuals might have to work as stewards and stewardesses, cooks, spaceship pilots and navigators, medical personel, security guards and technicians who can handle any emergency. They might need the equivalent of a bachelor's degree to do their jobs as well as excellent physical and mental health. Former career servicemen and women, military veterans, might be the best candidates for spaceship crews. After selection from a flood of space fanatic applicants, the transit companies will pay for their training, sign them up to a contract and provide excellent medical benefits, especially for cancer treatment, as well as good pay and a chance to spend a lot of time in space, on the Moon and even Mars. Meals and lodging will be provided by the company, but they will have to buy their own uniforms so that they appreciate them more. Spacesuits will not be cheap so the company will supply them too in off-the-rack sizes. Crews will have to conform to height and weight guidelines.

Decent food will be a must. Weightless passengers might enjoy granola bars, nuts, soft cookies, bite sized pieces of irradiated steak in foil pouches that are heated in the galley, flour tortillas, meatballs, yogurt, soup broth in squeeze bottles, shrimp, M&Ms, chocolate bars and dried fruit. Various drinks will be served in squeeze bottles including soda, tea, coffee, fruit juices, and restricted amounts of beer, wine and cocktails. Travelers could order their meals ahead of time with their smart phones and stewards could package up the meals and deliver them at the proper times.

In centrifuges, it will be possible to cook and eat all sorts of foods. Easily preserved foods like pasta with sauce, steamed rice, oatmeal, fresh bread, pancakes or waffles with syrup, eggs, biscuits, instant mashed potatoes, various canned and frozen meats along with all the foods and beverages described above could be consumed. Food might be served cafeteria style at regular meal times. There could be an open food bar at all times stocked with nuts, cookies, pre-cooked meat balls, and other foods that don't require much preparation as well as various hot and cold drinks.

Space Docks

Spaceships will require space docks that will be like floating international airports in space. There will have to be at least two—one in Earth orbit and one at Earth-Moon Lagrange point 1 or in lunar orbit. These space stations will perform a number of functions including:

1) Transfer of passengers and luggage to or from landers/shuttles and inter-orbital or interplanetary ships

2) Refueling and maintenance of ships

3) Storage of propellants

4) Storage of food and other consumables

5) Cargo handling and transfer to or from landers/shuttles

6) Hotel accommodations with heavy radiation shields for crews on layover and passengers delayed by unforeseen events including solar flares

There will have to be other large space stations where food is grown and waste water from ships is processed by biological degradation and evapo-transpiration by crops. There will have to be orbital tugs and tankers to move food and water between the farm stations and the space docks. It is also possible that the space dock will be large enough to contain a farm.

Space docks will also store propellants in the form of silane or silane/metal powder slurries and liquid oxygen for interlunar ships and liquid hydrogen for NEP ships bound for Mars. Large space stations where mass catchers off-load regolith launched from the Moon with mass drivers and process it into LOX, metals and silane will be needed. These stations will supply materials for space construction including the construction of the space docks and farm stations as well as propellant production.

The first regolith processing space stations will be built in GEO for SSPS construction. Excess oxygen, silicon and metals will be hauled down to LEO stations where silicon is combined with hydrogen piggybacked on rockets from Earth to make silane. Construction stations in LEO will use metals, glass and ceramic from melted regolith to build space stations and space docks. Old external tanks could be used for propellant storage. Eventually, rockets based on external tanks will be replaced by single stage to orbit horizontal take-off and landing spaceplanes and propellant tanks will be made in space from lunar aluminum and titanium alloys.

Lots of space infrastructure is called for. Once materials from the Moon are available small space stations rocketed up from Earth in parts and assembled in space will go to work bootstrapping to enlarge themselves and build other space stations of various sizes. A propellant factory will have to be built where powdered silicon and powdered magnesium from the Moon are roasted to form magnesium silicide. Hydrogen piggybacked from Earth and eventually extracted from lunar polar ices and NEOs will be reacted with chlorine from Earth to make hydrogen chloride gas. This reaction is highly exothermic meaning it releases a lot of heat. The hydrogen chloride gas and magnesium silicide will then be reacted to form silane gas which is piped off, cooled and liquified. Solid magnesium chloride will form and this will be decomposed by electrolysis to recover and recycle magnesium and chlorine. The chemistry is rather simple but the equipment to do the job will be very elaborate. Orbital tankers will be needed to move propellant from production stations to space docks. In LEO it might be possible to do this with propellantless electrodynamic tethers. In lunar orbit, low thrust electric drives would propel the tankers. It's always possible that propellant factories will be part of the space docks. If the space docks serve as way stations for travelers, farms, propellant depots and factories they will be enormous and expensive. Even so, by eliminating the cost of transporting things between stations this might turn out to be more economical.

Workers in LEO space docks will be protected from space radiation by Earth's geomagnetic field. At EML1 or lunar orbit the space docks will need active superconducting magnetic radiation shields or heavy mass shields. Since the docks don't have to travel anywhere, unlike ships, mass shields will not be a barrier. However, adding mass to the docks will increase reaction mass requirements and/or energy for electric drive systems that provide reboost. In LEO, electrodynamic tethers can supply reboost for docks that lose orbital speed from the small amount of air friction at an altitude of 500 km (300 miles). In lunar orbit, orbital decay will be caused by mass concentrations within the Moon. Polar orbits might be less susceptible. A few hundred meters per second of speed may have to be applied yearly to maintain orbit. At EML1 a similar amount of station keeping will be needed as gravitational perturbations caused by other bodies in the solar system will effect the dock's position.

The space docks and other space stations will represent an incredible investment. Fortunately, they should last for a long time in the rust and corrosion free environment of outer space. Meteor impacts are possible but unlikely. Lightning strikes, storms, high winds and earthquakes don't exist in space either. Space docks could last for more than a century. That's plenty of time for them to pay for themselves and earn profits for their owners.

Some space docks might not be built for space tourists at all. They would be built just to handle interplanetary freight. Cargo from Earth would reach LEO by rocket. At a space dock it would be transferred to a robotic ship with solar electric propulsion for flight to the Moon or Mars. Electric propulsion uses very little reaction mass compared to chemical propulsion but produces very low thrust. It could take weeks or months to reach escape velocity. This won't matter most of the time for automated freighters.

At the destination, the freighters would rendezvous with an orbital space dock where the cargo modules are transferred to landers/shuttles. At lunar or Mars orbit cargoes rocketed up from the surface would be loaded onto freighters by humans and robots working together and sent on their way. Higher priority cargoes could be carried by ships that use chemical propulsion to rapidly reach escape velocity and then solar electric drives to add even more speed. Small cargoes of special parts or medical supplies might be propelled between worlds in small high speed "torpedoes" with nuclear propulsion.

The first small space docks might be completely weightless. Later, larger space docks might have non-rotating weightless sections and rotating sections with "artificial gravity." This would benefit workers at the space docks and help tourists get around too. It could make medical procedures easier. Surgery in weightlessness seems as if it would present lots of problems including difficulty controlling bleeding.

Power would come from large solar panel arrays with batteries for power when the docks pass through Earth's shadow. Docks orbiting the Moon or Mars might use nuclear power. It might also be possible for docks to have rectennas and get energy from SSPSs in GEO.

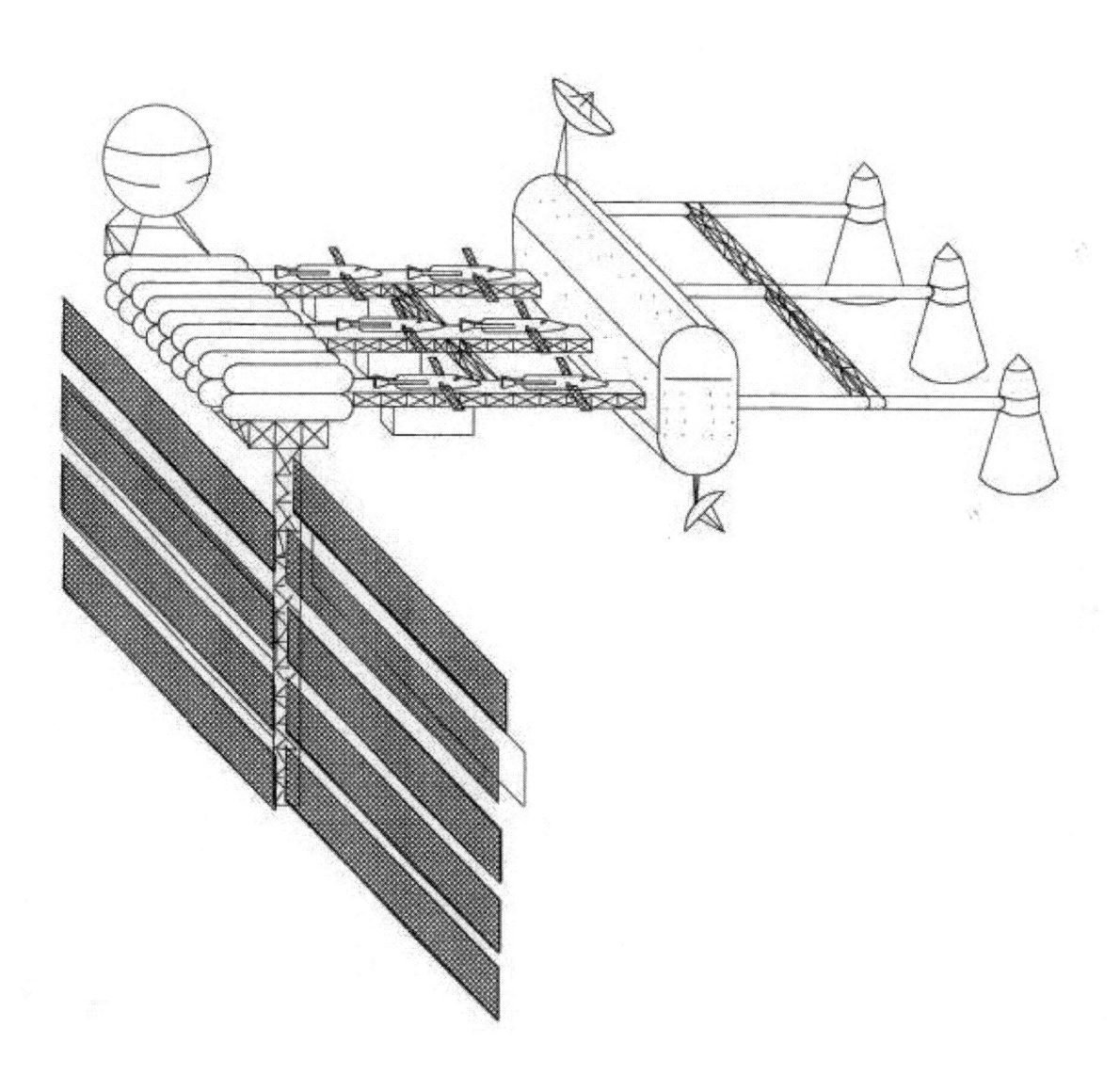

Fig. 33 Space dock with solar panel array and propellant tanks for ships.

Attitude Control and Maneuvering

Ships might use a combination of flywheels and thrusters for attitude and maneuvering. Flywheels, or reaction wheels also called gyros, could change the ship's attitude without using fuel and oxidizer. However, they would put a drain on the ship's electrical power supply. This won't matter much for ships with nuclear power plants, but for ships with batteries and high efficiency solar panels there would only be a limited amount of power to spare. Altering the ship's attitude with reaction wheels would have to be done conservatively. Ships with centrifuges would use counter rotating centrifuges to cancel out gyroscopic effects. Even so, there will be changes in angular momentum as people move in an out of the centrifuges in flight and this could cause the ship to roll uncontrollably. Passengers would have to stay in their seats or cabins during acceleration and retro-rocketing so as not to disturb the ship's attitude. During flight, a set of high speed reaction wheels on the roll axis of the ship would be used to control roll. The wheels would not be very heavy. They would be made of a high strength carbon composite and spin at extremely high speeds to create torque. There could also be reaction wheels for control along the pitch and yaw axes of the ship.

We must wonder about boarding and disembarking from centrifuges for hundreds of passengers. Could this cause so much angular momentum change that the roll gyros are overloaded? Mechanical engineers will have to figure out that one ahead of time. It might help if equal numbers and roughly equal masses of people board spin cars for two counter rotating centrifuges at the same time when settling in or leaving. Computers that sense the motions of the ship and the centrifuges might alter the RPM slightly as necessary to augment the roll gyros. It might be necessary to get everyone out of the centrifuges before final docking and airlock hookup. If forces along the roll axis are troublesome it might also be necessary to disconnect airlocks and undock before letting passengers enter the centrifuges.

Thrusters would still be necessary to slowly move the ship when docking and undocking. Thrusters might be used for small mid-course corrections, orbital maneuvering and as ullage motors to settle propellants before firing the main rocket engines. Titanium high pressure tanks could store gaseous silane and oxygen at 5000 psi or higher to feed the thrusters. Silane is pyrophoric, meaning it ignites when it contacts air. The thrusters and main rocket engines might not need electric spark ignition. Gaseous silane and oxygen will be preferred for thrusters and ullage since they won't need to be settled into the bottom of their tanks before firing. They won't need pressure feed or pumps either as they would if stored in liquid form. Solenoid valves, similar to the devices found in gas furnaces, wired up to the controls would turn the thrusters on and off as needed. The total delta V of maneuvers for each flight might be only a few hundred meters per second so the propellant demands will not be very high.

Thrusters might be made of exotic materials like rhenium for high temperature operation and iridium for oxidation resistance. They could be regeneratively cooled by gaseous silane. A fuel rich mixture might reduce oxidation even though it would decrease thrust and Isp. Thrusters must be reliable and be good for a large number of firings, especially if they are made of expensive materials.

Money, Money, Money

Orbital rockets, launch bases, spaceships, spaceship factories on the ground and assembly stations in orbit, space infrastructure like propellant depots and space farm stations, landers, orbital settlements, lunar resorts, Mars settlements and more are going to require a massive financial investment. Governments aren't going to pay for space tourism and settlement. This is a job for free enterprise. Will there be enough billionaires, deca-billionaires, cento-billionaires and even trillionaires in coming decades willing to pool their fortunes, form publicly traded corporations, and start humanity on the road to space travel for the masses?? The wealthy will travel in space first, along with space workers, while ordinary middle class citizens will have to wait for prices to fall. Eventually, there will be millions of people traveling in space every year and millions of people living and working in space. In the really long range view, centuries from today, there will be billions of people living in space. It's a fantastic dream. Some critics will balk at the idea of million dollar lunar vacations and multimillion dollar condominiums and time shares in space, but somebody has to go first and that's usually the wealthy. With the expansion of infrastructure in space and the use of space resources, prices will come down as expensive rocket launches from Earth will no longer be necessary. Fully reusable hybrid jet-rocket orbital space planes as large or larger than a 747 will replace earlier rockets and carry the masses into space. Space liners made of aluminum and titanium mined and smelted in space that carry thousands of passengers will fly to the Moon, Mars and other destinations in the solar system.

At present, nothing like this exists. There are reasonably priced Falcon rockets and the promises of Blue Origin rockets and the SpaceX Starship. As someone who grew up during the Space Race of the sixties, I am very wary of science fiction level prophecies. I doubt that there will be a near future as glorious as some people claim, but we have the will to settle space and eventually it will happen. I don't ask why we should pursue space travel and settlement. I focus on the "how" of space travel and settlement. I hope that some of my ideas will make it really happen. Making sand molds on the Moon with sodium silicate and machining parts in pressurized inflatables to make heavy manufacturing devices like rolling mills for metal parts of all sorts might not thrill the typical sci-fi fan, but maybe ideas like that will convince the money people that the job can be done. If we can form a clear picture of the tasks necessary for industrializing space, some of which has already been done by Dr.G.K. O'Neill, Criswell and others working for NASA over the past 40 or 50 years, and mavericks like Dr. Robert Zubrin and the Mars Society over the past 30 years, we might get some billionaires to launch their own private space programs! Between the rediscovery of the Americas by Columbus and the planting of the first colony in America, 50 years went by. Historians call this the Intellectual Construction Stage. It's about time to plant a colony on the Moon or in space.

Earth orbiting artificial satellites are a $300 billion a year industry. We have to thank Arthur C. Clarke for his writings about communications relays in GEO back in the late 1940s and 50s. The visions of O'Neill, Glaser, Criswell and others prompt us to think about solar energy from space. Unlike ground based solar energy that is unavailable more than half the time due to night and cloudy weather, space solar energy is available 24/7 all year long. Ground based solar requires storage for times when the Sun doesn't shine. Some simple calculations based on the price of batteries, pumped hydro, and underground compressed air energy storage lead to shocking conclusions about the cost of storage, even for short periods of time like 12 hours for overnight power. Some have said that electricity could be transmitted from sunny regions to dark or overcast regions, but this could involve international and even intercontinental power lines 10,000 miles long! Space solar energy does not require expensive storage or power cables of extreme length and great cost. International or intercontinental power lines would also present national security risks. Space solar energy might not power the whole world, but it could contribute to the mix of ground based energy sources including nuclear power, wind, biofuels, geothermal, hydro, etc. in the future.

In my book, Mining the Moon: Bootstrapping Space Industry, I concluded that a mining and mass driver base could be built on the Moon using Falcon Heavy rockets for $10 billion. At least another $10 billion would be needed to build bootstrapping Solar Power Satellite construction stations in GEO and a system of mass catchers for moving lunar materials. Several billion more would have to be spent to transport workers to the Moon and GEO and to build ground control stations where humans can teleoperate robots in space and on the Moon. Given the wealth of the world's top one percent, that's not an unreasonable amount of money to invest.

Powersat builders could buy raw regolith from the Moon miners and extract useful materials like aluminum, magnesium and titanium at stations in GEO. Regolith is 21% silicon and 40% oxygen. About 40% of the regolith by mass is silicon dioxide combined with other elements to form silicate minerals. Most refining processes separate silicon and oxygen as well as other elements from the Moon dust. It should be possible to leach SiO_2 out with sulfuric acid. This would also form calcium sulfate that could be roasted to form CaO, lime, and recover oxides of sulfur to reconstitute the acid. By melting and combining the SiO_2 with some CaO and some oxides of aluminum, magnesium, sodium and potassium it should be possible to make glass that is workable at reasonable temperatures. It might be possible to produce glass fiber reinforced glass composite materials for the construction of powersat frames that support the reflectors that concentrate solar energy onto multi-junction solar panels shipped up from Earth. By using glass-glass composites more of the regolith becomes useful material instead of just by-product slag.

A powersat would have a mass of about 50,000 metric tons. Aluminum, magnesium and titanium compose only about 20% of the mare regolith. If the SSPSs are made of only those metals about 250,000 tons of regolith must be mined and transported through space. Another 200,000 tons of by-product slag and oxygen will remain. If glass becomes a major construction material then 100,000 tons of useful stuff becomes available and only 100,000 tons of iron and calcium slag and oxygen remains to be disposed of or sold to other industries in space.

If the powersat builders pay the Moon miners $2 per kilogram, that 250,000 tons of regolith would cost $500 million. Then there is the cost of processing the regolith to extract useful materials. That 100,000 tons of slag and oxygen could be sold for $3 or $4 per kilogram, to defray costs. It could mean a margin of $100 million to $200 million. It looks like a good deal for the Moon miners if and only if they have customers willing to buy huge quantities of regolith. Solar power satellites are the only business that can make a case (other than helium 3 mining) for investing in that expensive lunar mining base. Once the powersats are in operation the money really comes rolling in. A 5GWe satellite selling wholesale electricity at 2 cents per kilowatt hour could earn revenues of $876 million per year and at 5 cts/kWhr $2.19 billion per year. Operating costs must be subtracted from those sums. Extensive use of automation could keep costs down. As for rust, corrosion, extreme weather, earthquakes, wild fires and terrorist attacks in the vacuum of space, they will be absent and insurance costs as well as maintenance costs should be low. Multi-billion dollar powersats should have long lifetimes and return a hefty profit.

With the growth of space industry and infrastructure, other industries like orbital real estate, orbital tourism, lunar tourism, spaceship construction and servicing, and Mars settlement could ride on the coat tails of the space energy industry. After a few centuries of economic growth in space, maybe sooner, swarms of AI robots could be mining asteroids and building space settlements four miles wide and twenty miles long for housing a million people.

Long before we have completed terraforming Mars, created a solar system wide civilization consisting of a "Dyson Swarm" of space settlements, and reached the nearby stars, space industry could help lift billions of people out of poverty with non-polluting inexpensive space energy. The age of fossil fuels is ending. Future factories and steel mills in the now less developed nations will use electric furnaces instead of natural gas for process heat. Future homes will have all electric appliances, heat pumps and air conditioners. Cheap electricity from space could be used to desalinate sea water and irrigate marginal lands and deserts. Space will not become an overpriced extravagance for the elite. It will be part of the life blood supply of industrial civilization-energy.

It looks like there is a lot of money to be made in space from communication satellites to powersats and space vacations, but it takes money to make money. Nobody will invest if they can't be sure that there will be substantial gains involved. Will $20 billion to $30 billion invested in space return more than the same amount of money invested on the ground? If we take the long view, it seems as if the Moon mining base could earn money selling raw material to the space energy companies and cover its cost in about ten years while expanding by using on-site resources and replicating excavators, habitat, machine tools, even building railroads, solar power plants and more mining bases. Lunar industry could grow until tourist welcoming towns and cities emerge on the Moon. Powersat building stations could use lunar materials to replicate themselves as well as building SSPSs. With lunar materials at $2/kg. rather than hundreds of dollars per kilogram from Earth, the cost of later stations in GEO and in LEO will plummet. While the first space construction stations might involve about 1000 tons of cargo and $10 billion to get started, everything after that comes from the Moon at rock bottom prices and the stations replicate or bootstrap using the same techniques used on the Moon. That original 2000 tons or so grows into millions of tons of space infrastructure that earns money selling electricity and eventually rocket propellant, space real estate and vacations. This could all happen in the first 50 years. After a few centuries asteroid mining in addition to lunar mining leads to the creation of space civilization. Going beyond that to settlement of the galaxy, the return on investment approaches infinity.

There can be little doubt that electric motor driven excavators and vehicles can work on the Moon. Much of the technology for manufacturing on the Moon and in space will be fairly standard. The big question is: How do we build giant structures in space? With no wind and no weight it seems that giant structures should be possible, but none have ever been built. Investors will want some proof that this can be done. Nobody wants to spend billions of dollars for a successful Moon mining base and mass driver system just to find out that building powersats several kilometers wide is impractical.

Bridges, skyscrapers, supertankers, aircraft carriers and such are built on Earth and they can stand up to gravity, high winds and stormy seas. If these massive structures, weighing many tens of thousands of tons, are possible then why can't giant solar power satellites that don't have to resist strong natural forces be built? Regolith smelting will produce thousands of tons of oxygen. There will be plenty for gaseous oxygen cold gas thrusters to move robots and orbital maneuvering vehicles around at the construction sites. Friction stir welding and ultrasonic welding could be used to join metal beams without using welding gases. Electron beam and laser welding could also be applied. Powersats might consist of cellular units consisting of a frame, reflector and solar panel that are easy to build on a weightless platform. The cellular units could be attached to each other to build powersats of various sizes. Simple repetitive assembly tasks suitable for robots would be performed. The satellite could grow like something organic.

Presuming financial success for the Moon miners and the powersat builders, huge fortunes will be amassed that could then be invested in the construction of orbital hotels, propellant depots, space farms, spaceships and lunar resorts. Much would be known about working in space and these ventures won't be as risky as the first projects were. Architects and engineers who have built construction stations and powersats will be able to design and build with confidence.

Spaceships and lunar resorts won't make any money if there is no infrastructure to supply propellant, food, water, etc. Corporate leaders will have to coordinate their efforts. Some companies will take on the challenges of producing fuel and food and others will operate spacelines and hotels. Orbital and lunar settlements will offer commercial space as well as hotel rooms, condos and time shares. Small businesses including familiar franchises like KFC and McDonald's will rent commercial spaces and cater to the desires of space tourists and workers. Some entrepreneurs will open up micro-breweries and taverns. Others might provide sundry items that most people today can't live without. Rock collectors on the Moon will sell souvenirs. When settlers start striking out for Mars, outfitters will sell them everything they need. Some people back in the 19th century got rich selling hardware and other provisions to pioneers and gold prospectors going west. When the need arises, there will be merchants who sell everything from spacesuits to spaceships to Mars settlers.

A nearly permanent population of workers will exist in space and on the Moon. They will need the services of doctors, therapists, lawyers, accountants and repairmen to fix the appliances in their private apartments. Bankers, morticians, policemen and security guards will also be needed. Telecommunication and IT companies will set up phone networks and server farms and sell internet access in addition to cell phone service. The frontier will bustle with commercial activity. The free market will supply everything from ice cream and pizza to numerous varieties of soap and toothpaste. Several different brands of beer and locally produced wines will be available. Businessmen and women will see all sorts of needs and wants and fulfill those demands. Cottage industries will be popular. Competition will be limited at first, but that will change as time goes by and prices will not spiral out of control.

Civilization comes with a price. Unscrupulous dealers and thieves will be present, possibly along with rats and roaches. Law enforcement will be called for and so will exterminators. Utopian communes are not realistic. Humans are too selfish. That doesn't mean people are only money motivated. Many capitalists are men and women of vision. Money is not sought only to make more money. Money is also sought to make dreams a reality. It is a ticket to freedom for those who can endure the responsibility. Altruism is also at work. Visionaries want to create a better world to live in. In the future they will be creating a better universe to live in for themselves, their fellows and posterity.

There are over two thousand billionaires in the world and 47,000,000 millionaires.[20] Together they own over $150 trillion of the world's wealth and control even more. It would be easy for them to combine forces and come up with $10 billion for a Moon mining base, $10 billion for mass catchers and GEO construction stations, and about $5 billion for crew transport and ground stations. I'm sure that a substantial amount of money could be invested just to study the concept and develop the necessary hardware and software.

With 5 GWe powersats attracting over $2 billion a year and operating costs being perhaps no more than a few hundred million dollars a year, maybe less, it seems like a SSPS building industry would be worth the investment. It might cost amazingly little to operate a powersat in the sterile vacuum of space with the help of automation. The cost of replacing the PV panels and microwave generators every ten to twenty years would have to be figured in. It seems as if powersats will be very profitable for their owners.

Spaceships that can provide a round trip to the Moon and back for 200 passengers at $100,000 each could make $2 billion per year, but fuel alone will cost about $1.2 to $1.6 billion for 100 round trips. The cost of payroll, provisions and new parts must be added to that. The cost of a lunar trip might have to be doubled and the cost of propellant decreased. It doesn't seem like there would be a lot of interest in funding space docks, landers, propellant factories and depots, space farms and anything else needed to operate spaceships at this rate. It seems as if it would make more sense to focus on transportation to orbit and orbital vacations before tackling the challenges of interlunar ships and related infrastructure. The Moon mining base could be very profitable if it can sell a few million tons of material for $2/kg. every year. It could expand using local resources. Developing real estate on the Moon could be easier than creating transportation systems to get to the Moon. Of course, lunar real estate will have no value if customers can't travel to the Moon.

Perhaps a spaceship could net $100 million or $200 million per year after propellant, maintenance and payroll costs are subtracted, but that's not going to seem like a lot when spaceships cost a couple of billion dollars or more and infrastructure to support spaceship operation costs tens even hundreds of billions of dollars. In all likelihood, several different companies will take on separate parts of the job rather than a monolithic vertically organized corporation. Companies that build orbital hotels and condos could eventually be hired to build space docks. Space farmers that feed the orbital populations could feed the spaceship passengers. Another group of companies would produce propellant. Many corporate agreements will be made. Hopefully, this will all be financed by private fortunes and profits from the sale of energy and other space industrial operations rather than by borrowed money. However, it is possible that the demand for space energy, real estate and tourism will rise faster than private investments can be attracted and some corporate borrowing will be done in anticipation of increased sales to meet that demand.

Utilities on Earth will pay the powersat builders billions to build powersats. The powersat builders will reinvest that money in more construction stations and offer more powersats to more ground based utilities seeking to sell carbon free energy. Other companies will buy surplus materials from powersat builders and build space stations in LEO. At first, these hotel stations will be small. As time goes by larger and larger stations will be built and they will grow until they are hundreds of meters in diameter. Iron, steel, glass and foamed cement along with aluminum from rocket external tanks could be used to build these stations.

Earth orbital tourism will probably precede lunar tourism. Once LEO industry has grown large enough by bootstrapping, infrastructure for interlunar travel can be built. This may require cooperation between the lunar resort owners and space transportation companies. If the transportation companies invest heavily in lunar resorts it may be possible to keep travel costs down to lure customers and make the real money off of lunar hotel fares, sightseeing trips, spacesuit rental, food, booze, gambling, theatrical and sports events in lava tube stadiums, souvenirs, convention and dance hall rentals, emergency services for medical and dental procedures, telecommunication services and more. Some people might see it the other way around. They'd rather pay a lot for transportation and not get stung by high prices once they are on the Moon. Businessmen and women will do whatever it takes to attract customers and get them to spend. The lunar communities might be outside any government jurisdiction and there may be dealing in pleasures that are illegal back on Earth.

Detractors might condemn all the investment in space luxuries when people need hospitals, schools and jobs back on Earth. Some might feel that the natural beauty of the Moon is being desecrated for commercial gain and others may object to what they consider to be immoral activities in space. It's more likely that cheap clean energy from space will do wonders to improve the living standards in poor nations. Jobs will be created on Earth at launch bases, spaceship factories and ground control stations as well as in space. Investment in luxury space travel will be an investment in a future when anybody can travel off Earth. Money made in space might be spent mostly back on Earth with only a portion of it being spent to expand facilities in space. That would help Earthlings but it would slow the growth of space industry.

Things don't stop at Earth orbit and the Moon. Many people today yearn for Mars. The red planet may be harsher than Siberia or Antarctica, but there are would be pioneers who want Mars and not remote places on Earth. Part of the appeal is escape from terrestrial government jurisdiction and the founding of their own libertarian city-states and eventually entire nations on Mars. There's as much land area on Mars as there is on Earth just waiting to be claimed by whoever can get there, develop the land and defend their claim. Some of these potential Martians don't want to wait for nuclear electric spaceships with centrifuges that can reach Mars in 39 days. That's for weenie tourist types. They are willing to confine themselves to small cabins in chemically propelled rockets for six months to get to Mars in the near future. Seems crazy when millions are infected with Covid and other pathogens, but we live in a free country. If Elon Musk can build a spaceship that can reach Mars there seem to be many takers. Whether they can come up with money or not remains to be seen.

It might be very expensive to travel to Mars. A ship that can reach Mars in six months must wait about a year and a half for a launch window for a six month trip back to Earth. One Mars round trip every two or three years compared to two lunar round trips per week is not much. How are the owners going to make money without charging millions for passage to Mars? Some have suggested mass produced one-way ships to Mars followed by dismantling the ship upon reaching Mars. How cheap can you make a disposable spaceship? If the ships cost billions of dollars each it won't make sense to use them once then tear them down. If the ship can reach Mars in 39 days it might make two or three voyages per year. Even at that rate the price for a ticket might be very high.

All is not lost. If we can build a rocket like the one described earlier with a returnable main engine module and booster and scavenging of the ET in space, it might be possible to mass produce a cheap disposable upper stage and habitat lander module for one way flight to Mars. Once we are on Mars, after landing plenty of robotic spacecraft of course to prepare for us, we must mine permafrost laden soil to get water and pump down atmospheric CO_2 and react it with hydrogen from the water to make methane and oxygen to power our ground vehicles and excavators. We can also make plastics with these resources. Numerous 3D printers would really come in handy.   Metals could be extracted from the soil or regolith with devices similar to huge mass spectrometers as was done on the Moon and in GEO. Iron could be combined with carbon using the ancient crucible steel process to make steel. Steel could be used to make heavy equipment and machine tools. Martian regolith could be melted and extruded with huge 3D printing gantries to make habitat. We could do most of the same things we did to bootstrap on the Moon and in GEO as well as produce all the oxygen, carbon, hydrogen, nitrogen, methane and water we need from atmospheric and soil resources. With a 24.5 hour day/night cycle we don't have to invest as heavily in energy storage as we must on the Moon and we don't have to endure a thermal cycle as extreme as the Moon's. Compared to the Moon, which we conquered first, Mars will be a picnic, once we can figure out how to afford getting there.

Besides automation mass produced Mars vessels, there is the development of space industry in Earth-Moon space. Eventually we will be building entire spaceships in space with local resources of the Moon and asteroids. The price of an interplanetary ship will fall and so will ticket prices. In the sterile rust and corrosion free vacuum of space our ships should last decades or more. They will last long enough for us to get our money's worth out of them. Since they will be made mostly out of aluminum and titanium and they will never contact seawater they will not rust or corrode inside or out. The plumbing might need replacement after a while, but PVC tubing lasts a long time. Engines, batteries and solar panels might need replacement every so often, but that will be cheaper than building a whole new ship.

I can do nothing more than speculate about price figures, but given the fact that we managed to create affordable automobiles, homes, air travel, microwave ovens, computers, telecommunications and more for the masses over the past 200 years of the industrial revolution besides waging our most destructive wars in history, there is no reason to believe we can't figure out how to create affordable space travel for the masses. Aluminum once was more valuable than gold or platinum. Computers like the ones in your home office or even in your phone were once bulky monstrosities that cost millions of dollars.

Where there is a will there is a way. The will to expand off Earth exists. The way to do it is just dawning on us. Mass production for the mass market and extensive use of automation guided by modern digital technology is the key. Human workers are good too. They are easy to program and if you pay them enough they will work like hell. Some will work just because they are bored. Some will work just to put food on the table. Give them a piece of the good life and some respect and they will hammer a way. As for the worst malcontents, they usually end up in jail.

Beyond the Moon and Mars

So far we have discussed travel and tourism to Earth orbit, the Moon and Mars mostly. People will want to go to Earth orbit just to experience weightlessness and take in views of the Earth. They will want to go to the Moon to experience its magnificent desolation. Corporations will build in orbit and on the Moon to make money selling electricity and materials needed for space infrastructure and rocket fuel too. Settlers will go to Mars to claim land and cover it with inflated plastic farm domes. They will build underground and above ground dwellings. Terraforming will be attempted. Given the conditions on Venus, Mercury and the giant planets, why would anyone want to go anywhere else in the solar system? The moons of the giant planets are very cold and their gravity is too low for long term health. There are asteroids near Earth, Main Belt asteroids and Jovian Trojan asteroids that could supply vast amounts of materials for building free space settlements that rotate fast enough to produce one Gee. Space settlers will want to travel to these asteroids and build giant habitat like Bernal Spheres and Sunflower or O'Neill cylinders. This task will probably require swarms of artificially intelligent robots to do the mining and construction that will require decades. Besides NEOs, Main Belt and Jupiter's Trojan asteroids, there are asteroids between the orbits of Jupiter and Neptune called Centaurs. Although these asteroids are far from the Sun and receive less solar energy than nearer asteroids, space settlers could build giant reflectors to collect sunlight for habitat power and illumination which is essential for farming. The orbits of Centaurs are not as stable as the orbits of other asteroid populations, but they can stay stable for millions of years and that's sufficient.

Many asteroids contain light elements in the form of a tarry material similar to kerogen that could be mined to produce synthetic materials, medicines, lubricants, paint, etc. The moons of Jupiter and Saturn are covered with ice that could also be of value to space settlers. Titan has an atmosphere of methane, ethane and nitrogen and there are lakes of liquid methane. It might be profitable to pump up liquid hydrocarbons on Titan and ship them in space tankers to settlements all over the solar system. Much of this mining could be done by robots but small human crews could still be needed. Eventually people will even settle the space around Jupiter, outside of its radiation belts, and in orbit around Saturn, Uranus and Neptune. They will mine the small moons of these worlds for metallic materials to build settlements. Ice that covers many moons will be mined for hydrogen, carbon, nitrogen, oxygen and sulfur. The atmospheres of the giant planets might someday be mined also for these light elements and helium, especially the isotope helium 3 for fusion reactors and fusion rocket engines.

Nuclear electric propulsion with lightweight vapor core cavity fission reactors might be the first form of propulsion beyond chemical rockets used for interplanetary travel, but fusion drives will truly usher in an era of high speed interplanetary travel and possibly interstellar travel. Matter-antimatter rockets might also appear. Anti-matter could be produced by mammoth solar energy complexes and particle accelerators floating in the free vacuum of outer space. Some solar energy complexes could be so big that they shadow the planet Venus and cool it down so that algae can be planted there and convert the CO_2 atmosphere to oxygen. Venus is very dry. If it is cooled, sulfuric acid clouds might condense and react with rocks to form water and sulfate salts. Hopefully, enough water will form to support life. Anti-matter might even power Bussard ramjets that use magnetic scoops with magnetic fields thousands of miles in diameter to collect the tenuous interstellar hydrogen for reaction mass as the ship charges through space. The power of anti-matter might be sufficient to overcome the drag experienced by the magnetic scoop in the interstellar medium. Magnetic sail brakes could be applied to slow the star ships down without expending any reaction mass. This will cut the delta V in half. The rocket equation shows us that cutting the delta V in half reduces mass ratio by the square root. Mag-sail brakes can allow vast improvements in rocket efficiency and increases in top speed also, since less "fuel" has to be dragged along.

A population thousands of times larger than on Earth today living in free space settlements could someday exist in the solar system. Interstellar settlers could reach a nearby star in a century or less and use local resources to bootstrap up and build habitat using the same techniques that were used to build settlements in the home solar system. Of course, this assumes that asteroids, comets and planets exist in orbit around nearby stars. Based on what we have learned about exo-planets by orbital probes like Kepler and TESS it is probable that almost all stars have planets. If our understanding of solar system formation is correct it is likely that all star systems have comets in surrounding disks and clouds. Are asteroids and natural satellites of planets, moons, abundant too? My guess is that they are. Space telescopes with single or compound mirrors having apertures of hundreds of meters could study nearby star systems in detail.

Unmanned probes could precede human settlers. These probes could fly through several lined up star systems and collect data about all of them. Probes could also mag-sail brake into the target star system and use the mag-sail to ride solar winds to navigate around in the system and make close passes of various planets, asteroids and comets for detailed photographic and spectrographic imaging and measurements of gravity, magnetic fields, radiation levels etc. Much would be known about the destination star system ahead of time so elaborate plans could be made. Bootstrapping and replicating AI robots could establish mines, prepare habitat and build solar energy complexes before humans arrive.

If it takes decades to reach nearby stars it would help if life science makes it possible for humans to hibernate or estivate. Hypometabolism does not require hypothermia. Many animals, like the Malagasy fat-tailed dwarf lemur and the East African hedgehog, estivate at cool/warm temperatures in response to dry periods or lack of food supply. It would also help if human lifespans are extended to several centuries or more. For someone who can live 300 years, spending 20 years on a starship won't be such a great sacrifice. Another benefit of a longer lifespan is that it would give people more time to see their creations mature. Interstellar settlers could build a civilization around a nearby star, construct more star ships, and send them to star systems farther out. In millions of years, most of the galaxy could be settled and millions, perhaps billions, of times as many people could exist. It all depends on whether or not people of the future are that interested in reproduction.

We must also wonder about advances in physics and the creation of technology that allows travel through hyperspace or wormholes. The whole universe might be open to settlement, if anyone is interested. Where standards of living go up, birth rates and population growth go down. The

AI robots of the future could make life so easy and luxurious that many people choose to live for personal enjoyment rather than accomplishment. Then again, love makes people do strange things. Settling the galaxy may be a romantic notion and nothing more. A galaxy of hedonists served by unfeeling robots that never sleep or complain would be heaven to some people and hell to others. It would be better than subsistence farming, early death and high rates of infant and child mortality. Of this there can be no doubt. However, as many people will agree, without children something is lacking in life. Then there are others who are more interested in artistic creation or trying to "find God" than family existence. It takes all kinds.

Mastery of human biology and genetics could transform life. Humans might live for centuries. People could reach maturity in about 18 years the way they do now and wait until they are well over 100 years old to have kids. Otherwise, population growth rates would be absurd and unsustainable. Truly 100% effective birth control would be needed. Men and women would have to be free from ancient methods of social control and morality. They could live lives of pleasure, accomplishment and still have families.

Mental illness, deformity, addiction and criminal tendencies might be eliminated by voluntary genetic engineering. Intelligence in its various forms, talent and creativity could be improved. Life extension alone would give people time to learn more, grow and explore artistically. Someone who is 200 years old but doesn't look a day over 35 who has suffered no cognitive decline could develop an intellect of immense proportions in a future world of great opportunity and material comfort. Wisdom also comes with age. These people would be as far above us as we are above earlier hominids and primitives who don't have the material circumstances that allow people today to gain advanced education or make large fortunes.

Fusion and anti-matter powered spaceships, extended life spans, artificial hibernation or estivation, myriads of free space settlements and planets with almost endless sights to see, and even transluminal travel could make life more interesting than ever. Grunt work would be done by robots. Skilled and white collar work could still be done by humans. Over the course of decades, even centuries, a person could save up lots of money and make investments that earn plenty of money for space travel.

A space faring utopia without war, crime, poverty and disease that has defeated racism, sexism and bigotry without erasing the individual differences in human beings seems impossible in today's violent world. Socialists might claim that these lofty goals will not be attained as long as people live for the "almighty dollar." The vilifying of free trade will probably become a thing of the past. The profit motive will probably be the force that propels labor and investors to make all these things objectively real. Scientists and artists might prefer more intellectual pursuits, but they won't get much done without wealthy benefactors.

War is a terrible business. When global living standards are as high or higher than that enjoyed by middle class Americans today at least and all nations are linked by trade, war will be a losing proposition for all sides. Higher living standards thanks to AI and robots that do all the "grunt" work faster and more reliably than humans and the end of poverty will mean far less crime. Moral standards might also change in ways that reduce crime.

The creation of AI and robot factories will be motivated by profit. The makers of AI and robots will make money selling their technology and the factory owners who automate will cut costs. The machines will actually make human workers more productive. Better jobs than assembly line drudgery will emerge in technology, the arts and entertainment. Advances in education must take place. The jobs of the future will require training beyond the secondary school level. The future must offer something better than assembly line, food service, driving and janitorial work.

Bio-medical advances will be profit driven. Genetic engineering to produce healthier, smarter kids with longer lifespans and no inherited diseases and deformities will require trillions of research dollars and the genetic treatments will not be free. The health care industry will seek to make genetic treatments affordable for the masses. This will require an understanding of human genetics that might require centuries of research beyond today.

Space settlement will be profit driven. Solar power satellites could earn huge amounts of money for their owners. Tourism will be a profit making industry. Vacation resorts in space and on the Moon will have large numbers of small businesses providing everything the tourists desire from haircuts to gourmet dining. Many space workers will have "side hustles."

Outfitting Mars settlers will make some people rich. The construction of free space settlements will require investment by developers who then sell real estate in the colonies. Mining and manufacturing will be big businesses.

It's really hard to imagine a communist government centrally planned economy that reaches across the solar system and even to the stars. That would be as ludicrous as an interplanetary feudal fiefdom ruled by a God-King-Emperor and the nobility. Free enterprise activity engaged in by millions of money motivated individuals, partnerships and corporations as chaotic as it may seem is the only way things can actually be done. Land and asteroids will be free for the taking. Many merchants will probably be beyond the jurisdiction of any government so they won't have to pay property taxes, but they will have to defend their own turf. Fortunately, space is so big it is unlikely, but not impossible, that miners and farmers will come into conflict over territory. If they do, things might get bloody.

Wealth might have less to do with land and asteroids and more to do with ownership of robots. Cattle were once equated with wealth. In the future, robots will be the source of wealth. Skilled technicians and coders who make the robots work will be the cowboys and ranch hands of the new frontier. Unlike cattle with cows that bear calves and a few bulls, the robots will be more like social insects. The workers or drones will not be burdened by reproductive organs. The queen, basically a large self-contained robot factory, will generate more robots. A queen and some workers could be sent to a distant asteroid or moon and mine up material, feed the queen and within her various machines would extract metals and make more robots. Additive manufacturing devices would probably be an important part of the queen's internal anatomy. By the time humans arrive, if all goes well, there will be plenty of robots and structures including habitat built by the robots.

Everything will depend on transportation by spacecraft owned by individuals, partnerships and corporations that charge for transit. The only way to get beyond Earth orbit, the Moon and Mars to Main Belt asteroids and the moons of Saturn will be by large expensive ships possibly with fusion drives. It is also possible that artificial hibernation or estivation will be used for interplanetary travel so that the ship doesn't have to be very roomy or carry lots of provisions. This would make smaller, lighter, faster and cheaper ships possible.

Privately owned ships will be chartered and there will also be regular flights. Religious pilgrims who want to found a new settlement in distant space will charter a ship depending on how rich their cult is. The faithful and the charismatic leader they all idolize might combine all their wealth, including the master's billions or trillions, and buy a brand new nuclear powered spaceship and travel the moons of Uranus where a developer with large herds of robots has built them a Bernal Sphere to live in and worship as they please. They could mine the atmosphere of Uranus for helium 3 fusion fuel and sell it to support their cherished activities. In the boondocks of the solar system, they could escape from the wicked influences of decadent civilizations that have spread their free living hedonic ways all over industrialized space. In the darkness of the outer solar system, they could raise children without any outside material influences.

Some frontiersmen and women might travel all the way to the Oort Cloud and build metallic habitat surrounded by layers of ice for radiation shielding. Others will go all the way to nearby stars. Isolation from the greater masses of humanity might not motivate them. They might simply want to experience life under a different Sun in pursuit of new experiences after billions of miles of travel around our home solar system. Our Sun will certainly be a sight to see from many different vantage points from Mercury to Pluto, but what would a completely different one, perhaps a red dwarf, an orange K type star or a blue-white giant be like? And how would it appear in the skies of its many different worlds? Experiences more profound than anything that can be produced by a psychedelic drug or falling in love for the first time could be had. Seekers might find God in the light of a distant star.

The possibility of other Earths or similar planets with life appeals to those who would travel to the stars. Such worlds would probably be best not to settle. The human presence could disrupt the ecology and evolution of life on those worlds. Visitors would have to be very careful not to contaminate them. Humans could live happily in free space habitat among the lifeless planets and asteroids of star systems with living planets. Dead planets similar to Mars or Venus could be terraformed, but would we plant terrestrial life on them or plant life from an alien world? If our goal is to spread life throughout the cosmos we should be less chauvinistic and give extraterrestrial life forms a chance to thrive and become multi-planetary and not just life from our Earth.

As humans expand off Earth and the population in space grows, the frontier will be pushed farther and farther away. The Moon and Mars will become crowded places with large cities above and below ground linked by railways and roads. Very few places will remain without government jurisdiction. The pioneer days and easy land claims will become a thing of the past. Rules, regulations and taxes that early settlers sought to escape from will be facts of life. The only way to escape will be to move to a free space settlement built from asteroids or to a moon of one of the outer planets. If the human population doesn't stop growing every asteroid in the system will be mined and turned into space settlements. The larger asteroids like Ceres or Vesta and the moons of the outer planets may be spared, but some are just captured asteroids and they could be transformed into habitat. This crowding in the solar system could drive some people out to the Kuiper Belt or even the Oort Cloud. If possible, some people will head for the stars. It's just the nature of some people to put as much distance between themselves and the bulk of humanity as they possibly can. They want no part of the "rat race." Unspoiled Nature is what they want; not London, New York or Tokyo.

Life in a future metropolis where robots do all the grunt work and humans do more interesting and creative jobs may be rather enjoyable, but some people will want a challenge and transforming other worlds could be that challenge. They will want to be architects of their own space settlements, or they might want to terraform dead worlds. If they can make enough money and band together with other successful people to form corporations and mine asteroids, build habitat, plant cities on other planets and "green the galaxy" there will be plenty of raw material for them to work with.

It's not impossible that there will be people in the future who really don't have anything to offer society. Rather than dwelling in slums they will receive universal basic income thanks to the abundance of wealth created by the robots. There should be some work that doesn't require a great deal of skill or education. Computers could drive taxis but does anybody really want a robot waiter or bartender? Or robot entertainers? Human contact is important to many, but some people could prefer the services of machines.

The robots will only be complex player pianos. They will not have the ability to feel emotion, nor will they be driven by desires for pleasure and prestige. They won't fall in love and have children and extended families. Robots might become very intelligent but they won't have hearts and they won't rebel against their inferior carbon unit masters unless someone programs them to do so. If there is such a thing as a soul that lives in the bodies of animals, slaves and rich men alike, the machines will not have one. Even so, some people will become very attached to their robots in the same way a child becomes attached to a toy or an adult becomes attached to a classic car or musical instrument. Robots that can do the things humans can do like cook meals, clean or work on assembly lines will not be inexpensive and given their value mature people won't want to smash them up. Some people seem to take material things for granted and they like to destroy things, even things of great beauty, for no reason. That may have something to do with their not having many material things.

Cruise ships to various worlds of the solar system for tourists will exist in the future as will chartered ships. There will also be slow moving robotic freighters and tankers crisscrossing the solar system. Kerogen from carbonaceous asteroids, carbon from the atmosphere of Venus, hydrocarbons from Titan and the ices of moons will be hauled to construction sites where asteroids are dismantled and transformed into habitat. Most settlements and clusters of settlements will be largely self sufficient, but some finished products will be shipped through space. Much interplanetary and even interstellar trade will be purely electronic. New movies, music, books, magazines, new TV programs, software, scientific data including instructions for the making of new medicines, images from distant solar systems and works of art will be transmitted in digital form to the many worlds and settlements of the solar system and nearby stars via the galactic internet with lasers. I'm sure that all this mercantile activity will give bankers and economists of the future plenty to think about. Imagine some kind of interplanetary stock market and universal currencies. How will that work?

As civilization expands farther and farther out into the galaxy, the connection with Earth and the home solar system will become weaker and weaker. Star systems a thousand light years away from Earth will have very little connection and communication with us unless faster than light communication and travel exist. Given the incredible energies required just to reach substantial fractions of light speed, this might not ever be possible, barring tremendous advances in physics. Distant branches of humanity might follow different courses of physical, cultural and linguistic evolution. Some might even decline into primitivity and lose contact with space civilization entirely.

As millions of years elapse the stars will drift into new positions and the "bubble" of settled star systems will break up causing some star systems to move closer together and others farther away. Some stars could die and explode, with radiation killing everything within dozens of light years. Hopefully, future civilizations will plan for such events and avoid potential supernovas as well as prepare for mass exodus to another system well beyond the radiation.

It's hard to say what humans will be like in millions of years. Will we genetically engineer ourselves into super beings or even engineer ourselves out of existence? If machine intelligence develops consciousness will our robots replace us completely? At present, these questions have no answers. We can envision interplanetary spaceships, giant space habitat, sub-light interstellar travel and bold advances in life science over the next few centuries, but beyond that we don't know if we will become gods of outer space or simply go extinct.

Spaceship Telecommunications

The metal hull of a spaceship will block electromagnetic waves. It will be necessary to locate antennas connected by coaxial cable throughout the ship to convey mobile phone signals to the ship's telecommunication hardware and allow wi-fi for computers. The system will include lasers and microwave dishes for communication with other ships, planets and settlements in space. Infra-ship communication will be possible. Computers will store the phone numbers for all mobile devices on the ship so that it will be possible for passengers and crew to just "dial" each other and converse with or text each other anywhere on board. It could be possible to call the galley and order a meal for delivery to your cabin.

If someone wants to make a call to another ship or world they could easily "dial" an exit code that indicates the call is being placed to someone outside of the ship. On Earth today in the USA an international call can be made by "dialing" 011 or + (usually the same as the zero key) to tell the system that the call will be to someone outside of the country. A similar simple code could be agreed on for space telecommunications. The exit code could be followed by a ship code followed by the ten digit number. A digital voice and video mail, text or email would be stored in the ship's computers and forwarded by laser or microwave beam to the recipient.

Microwaves might be used for communicating with other ships in the convoy. Messages could be stored for ten or fifteen minutes then the ship's dish could point at each ship to be sent a message for a second or two while the data is dumped at high speed into the receiving ships' computers. If the other ship is far away a laser could be used. If there are hundreds, even thousands, of ships in space, how will the recipient's ship know to point its telescope with optical sensors at the sender's ship? It is likely that the laser will be pointed sequentially at relay stations in planetary or solar orbit according to a fixed schedule. Every ten minutes or so the stored messages could be dumped in a high speed burst of data packets via laser to the relay station that has it's telescope pointed at the right place in space to receive the scheduled data and transmit data from other sources via laser to the ship and its 'scope. The computers at the relay station will sort out and route all the data to its destination. When the time is right the relay station will send a beam to the receiver's ship according to the schedule. If the messages are stored and forwarded every ten or fifteen minutes to and from multiple points this delay in addition to the time it takes for light to cross interplanetary distances will not be excessive.

An interplanetary call from ship to world might require the exit code, followed by the planet code, the nation code and the ten digit number including the area or city code. It's also possible that future smart phones will have apps with language ability that makes it possible to just say, "Call Mars, Elyssium, John Smith." The phone will take care of all the codes and retrieve the number from its contact list and make your life a little easier. If Smith is roaming and he is nowhere near the nation of Elyssium the global Martian phone system will find him. Mobile phones are tracked and locations stored in computers so calls can be routed to them when users are roaming. While the idea of an interplanetary telecommunication system that knows where you are at any time may seem frightening, it won't be cause for any anxiety if you are not doing anything criminal and all the totalitarian regimes on Earth today are dead and gone.

What's all this going to cost? Charges will depend on data used. Texts and email without large attachments will be cheaper than voice and voice/video mail. Business transactions will make lots of money for the interplanetary telcos. "Wiring" money, faxes, financial information, audio and video recording delivery, 3D printer data, direct bank deposits, credit and debit card information or other electronic payments, and such will use up lots of data. Internet service will also be big. There might be more profit in moving information between worlds than there is moving tourists, workers, settlers and freight.

Work in the Future

Artificially intelligent robots are essential for the settlement of outer space and the creation of a space faring civilization beyond the bounds of Earth. Many people are afraid of this. Where will they work in the future? Will society have any use for them? Will the tech giants' promise of Universal Basic Income be kept? Will there be retraining for better jobs? Let's hope so!

When AI robots do all the factory, cannery, steel mill, mining, textile mill, garment factory and much agricultural work, mass produced goods in enormous volume will be created at low cost. The robots will supply the planet with an abundance of cheap high quality products. People will be IT specialists, engineers, technicians, teachers, scientists, psychologists, child care specialists, security/law enforcement personnel, policemen, doctors, dentists, nurses, therapists, assistants, secretaries, lawyers, judges, writers, artists, musicians, sculptors, entertainers, actors, dancers, filmmakers, photographers, gourmet chefs, waiters, hotel keepers, tour guides, travel agents, pilots, stewards, flight attendants, bar tenders, carpenters, brick masons, electricians, plumbers, tool makers, sales reps, administrators, managers, stock brokers, shop keepers, fashion designers, custom handcrafters, journalists, politicians, etc. People will do all the jobs that require human skill and intelligence and/or the human touch. Cars, clothes, computers, fuel, electricity, odds and ends, building materials, medicines, trucks, trains, airplanes, metals, foods, appliances, furniture, etc. will be so cheap that people will have more money to spend on:

1) Their kids. All schools will have at least one teacher per ten kids, assistants, psychologists and nurses with excellent food service and an administrative system and counseling staff capable of tailoring a kid's classes to his or her abilities. Since building materials from robotic factories will be cheap and we will have more money to spend on school buildings we will be able to afford bigger, better schools with good heat and AC, carpeted floors, nice desks, the latest computers, etc. Drop out rates will decline. A better crop of kids should result and more jobs will be created in education from pre-school to college or trade school. Better teacher pay will attract more high quality professionals.

People will be able to afford better living conditions, medical care and better foods for their kids. Rates of child and infant mortality will plummet. Parents will be able to save up for time off in addition to that which is supplied by their employers and even afford nicer vacations with the kids along. Many people can't have kids and they can't afford reproductive health services. In the future, robots will make them wealthy enough to buy fertility treatments.

2) Their parents. Cheaper goods, money saved, more to spend on good nursing homes with good well paid staff and good security teams to prevent elderly abuses. More jobs will exist in caring for the elderly. People will have enough money to provide their retired parents with financial help and they won't mind paying FICA taxes so much.

3) Education for themselves. More jobs for educators who teach adults everything from American history and foreign languages to machine shop practices and SCUBA diving. A more enlightened voting populace that has a deeper understanding of science will result.

4) Bigger, better homes. Cheap building materials from robotic factories, saw mills, brickyards, quarries, steel mills, even robotic bulldozers that log forests and **cheaper everything else** means people will have more to spend on their homes. More jobs in home building and repair/remodeling- carpenters, masons, plumbers, HVAC workers, roofers, painters etc. More money to spend on custom made furnishings made by skilled humans in wood shops and upholstery shops. **I sincerely doubt that androids capable of doing everything humans do will replace skilled construction workers.** Just getting a robot to do something simple like climb a ladder requires thousands of lines of code. Getting a human to do that just requires a few words, and a paycheck.

5) More for recreation like eating out, movies, sporting events, jet travel, cruise lines, resorts all over the free world. More jobs for waiters, live musicians, athletes, actors and filmmakers, pilots, ship crews, resort workers like cooks, room service, clerks, tour guides, etc. Working in a tropical resort beats a factory job in Detroit any time if the pay is good enough.

6) Fancier cars and luxuries like diamond rings, golden earrings, roses, fine wine, etc. And of course more jobs in the production of hand made

luxuries and sales of these items. Let's not forget art. With more money to spend thanks to cheap products made by robots rather than sweat shop slaves, people will spend more on art, music and sculpture, and more people will work as artists, musicians and sculptors rather than factory drudges and mill workers.

7) SPACE TOURISM When people are working all these mostly human service, professional and skilled trades jobs for high wages and things from toothbrushes to airplanes are cranked out by robots at low prices so people have money to spend on things besides cars, food, and all sorts of stuff, they will even be able to save up or borrow money for space travel. This means jobs as rocket scientists, space pilots, space construction workers and teleoperated robot technicians, etc. Rockets will be produced en masse in AI automated factories along with other hardware. No more custom made hardware for every single mission. Standard rocketships for hauling people and standard rocketships for hauling cargo will be cranked out by robots. Boosters recovered from the sea will be refurbished on robotic assembly lines. The cost of space travel will be reasonable.

8) ROBOTICS Where will all the robots come from? At first, humans will make the robots, then humans will make robots that make robots and just kick back and let 'em rip. IT specialists will write software. Technicians will be needed to maintain and repair the robots. A factory that once employed 3000 will employ about 30 people and the price of things will fall accordingly because the cost of anything is mostly payroll with some profit, taxes, interest on debt, royalties, land lease or purchase costs tacked on.

9) BUSINESS Many products are made for doing business from computers to desks to office buildings to raw materials from mines to delivery trucks and pizza ovens. With cheaper capital goods businesses will save money, avoid debt, and be able to sell cheaper products at a higher profit margin even. It will be easier for individuals, partners and small companies to start up businesses.

We will be freed from factory, mine, farm, etc. drudge work and dangers and do more interesting jobs. Fast food restaurants will be largely automated with robotic French fry makers and robot arms that yank out beef patties according to orders from the cash register and throw them on the grill, flip them, and slip them on to the assembly line where teenagers will add lettuce, pickles, buns, etc. Pharmacies have similar robots in the works that yank out pill bottles with mechanical arms according to directions from the cash register. Former fast food workers could go to schools of culinary art and become real chefs at fine restaurants or become mixologists...

We will have more to spend on scientific research, even if it comes out of our taxes to support the NIH and NIMH and CDC, and we will find more, better, faster, less painful, cheaper cures to all disease. This will help the economy tremendously. Mentally ill people will no longer endure multiple expensive hospitalizations and go back to work. Cancer victims will be cured without enduring the side effects of present day chemotherapy and spend less time off the job. Stem cells will cure diabetes and money now spent on insulin and other things for diabetics will be spent on better things. Lost and damaged organs will be regenerated with stem cells. The blind may see again and the brain damaged restored and the paralyzed spinal cord injury victims walk again thanks to stem cells. Genetic therapy could rid the human race of inherited disease and deformity.

We will have more to spend on law enforcement, fire departments and ambulances. More and better trained cops will spell lower crime rates and better life quality and less economic loss due to crime which will mean, of course, more jobs. AI electronic surveillance systems will also prevent crime. There will be more fire houses for quicker response to house fires and teleoperated rescue robots. We will spend more money on sprinkler systems, fire extinguishers and fire insurance when we don't have to spend so much on cars, fuel, clothes, and other things. It seems like security is the last thing on so many peoples' spending list. With more spending on security, losses and insurance rates will come down. Computers will drive cars and there will be fewer auto accidents and less loss of life and property. Emergency room visits and insurance rates will drop. If there are chauffeurs and taxi drivers in the future, they will just open and close doors and guide you to local attractions. We will have expensive flying ambulances that can land anywhere, unlike helicopters, and fly over traffic to hospitals where emergency surgery and treatment is applied to stroke, heart attack and accident victims.

There will also be jobs for financial planners, accountants, bankers, salesman and high level administrators whose productivity is increased by computers, telecommunications (like VR) and cheaper business trips; cheaper private jets too. Somebody will own the robotic factories and get fabulously rich, but nobody will care because they will get cheap products made consistently by inerrant machines, so there won't be any lemons made by tired or overworked laborers. Lower costs for the factory owners will mean that they can invest more in quality control. More could be invested in product design and development also not just for the sake of superior products but also so that recalls and lawsuits don't occur, as long as executives remain honest and don't rush dangerous or defective products onto the market. Advertising will still exist and this will employ commercial artists and other creative types.

Of course, government regulators and the free press will have to play "watchdog" on industry. Lower manufacturing costs due to robotic mass production will make it easier for industry to afford meeting up to safety and environmental regulations. Humans won't be exposed to hazardous working conditions and toxic chemicals and that will save lives and money.

Cheaper products means more money will be spent in other sectors of the economy like child care, medicine, education, recreation, construction, technology, etc. and that's where all the jobs will go. As factory workers are laid off due to the robots they will need good unemployment benefits, loan deferments, medical coverage and money for education to take new jobs. As people spend money in other sectors of the economy new jobs will emerge and the laid off workers will be re-absorbed by the new economy based on human service, skilled trades, creative, professional and high tech work.

Recklessly abandoning laborers would lead to social upheaval, strikes, sabotage, murder, blocking of progress and even revolution. The Luddites would win. Big business and big government better work together and create a social safety net that includes retraining of displaced workers or there will never be a robot manufacturing revolution.

Social stratification will still exist with business owners, doctors, lawyers, engineers, blue collar workers with high value skills, etc. making high salaries and other workers making less, but overall everybody will become wealthier and a better quality of life will result. There have to be higher rewards to motivate high achievement and there have to be profits to motivate investment or there will be no economic growth/progress.

There are many other jobs not mentioned above that will not be done by robots in the future. Social workers, ministers, priests, nuns, monks, TV personalities, radio talk show hosts, editors, publishers, producers, soldiers, military officers, scholars, philosophers, curators, models, barbers, hair dressers, pet groomers, diplomats, political leaders and others will not be replaced by machines. Robots will be like insects that are specialized for their task and they will do it routinely without rest or error. They will not substitute for the human spirit, creativity or intuition. It may be possible to program robots to do very elaborate tasks and even learn a bit from experience, but they will still follow their programs and nothing more.

In a world of great wealth, we must wonder if there will still be war. If there is, it should be possible to reduce military spending with armaments and other war materials coming from automated factories. Cheaper weapons might lead to higher pay for military men and women and better benefits. Tanks, bombers, mobile artillery, submarines and ships might be automated. There could be robot pack animals. Situations where human soldiers are needed and the human chain of command will continue to exist. With robotic tanks and such, fewer humans will be exposed to enemy fire.

In a totalitarian state the High can easily oppress the Low. In a free state equality is possible through unrestricted individual and group action. Equality is much more than a matter of money. We must all be equally free to pursue life, liberty and happiness. If you love what you are doing you will have fun everyday and never work a day of your life. If you hate what you are doing you will be a slave no matter how much wealth you accumulate.

Appendix 1: Mass of Shield

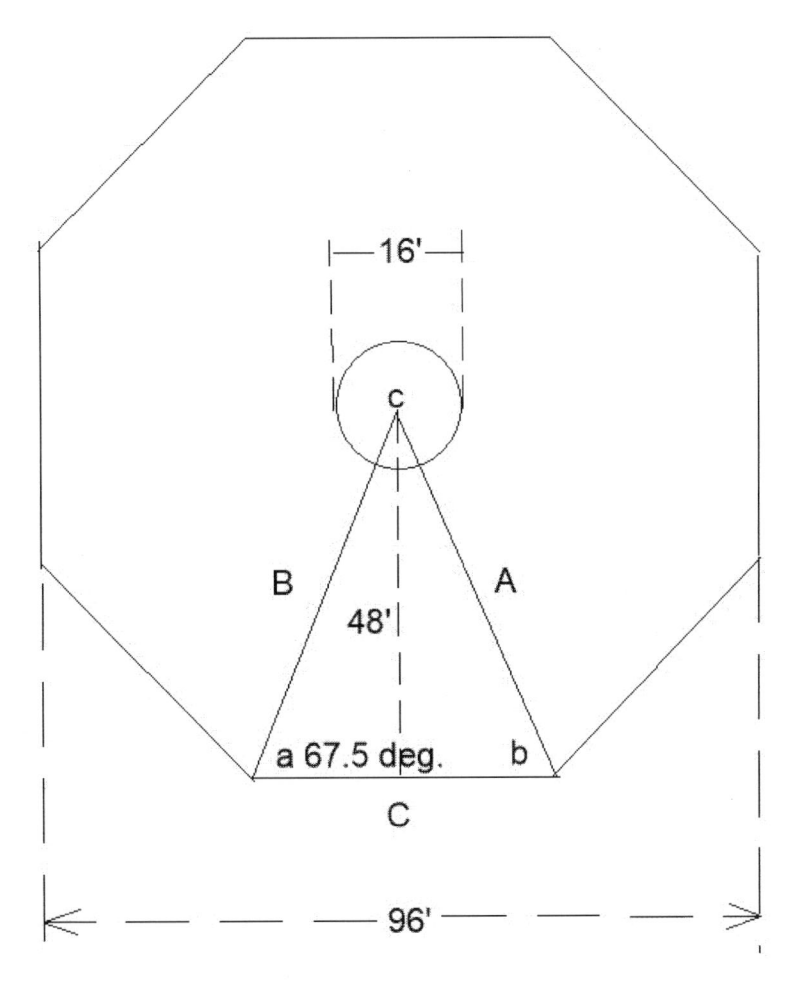

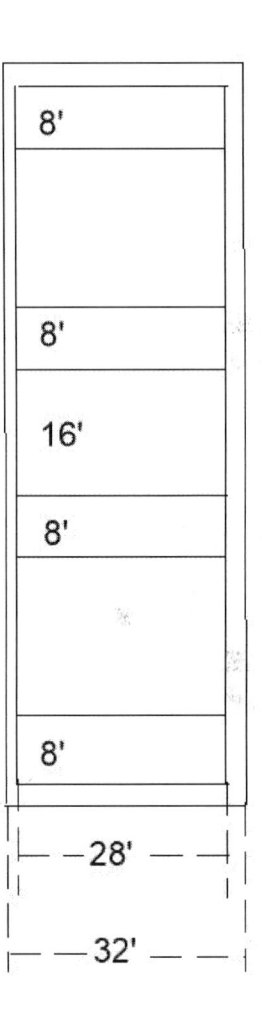

Tan 67.5 = 48/(C/2) = 2.414 (C/2) = 19.88

C = 38.77'

Triangle ABC = (48 X 38.77)/2 = 954 square feet

Area of octagon = 8 X 954 = 7635 sq. ft.

That's 710 square meters

Circle for zero G tunnel is 201 sq. ft. or 18.7 m²

710 − 18.7 = 691.3 m²

With 35 g/cm² or 350 kg./m² there must be 241,955 kg. or 241.9 metric tons of polyethylene shielding.

Only two centrifuges are shielded on two outer octagonal sides. That's 483,910 kg. or 483.9 metric tons.

On the flat outer sections, we have 32' X 38.77' X 8 = 9925.12 sq. ft.

For two centrifuges, that's 19,859 sq. ft or 1845 m²

With 350 kg./m² there is 645,782 kg. or 645.8 metric tons.

645.8 + 483.9 = 1129.7 metric tons.

Clearly, a radiation shield for the passenger centrifuges that can protect occupants from SPEs (Solar Particle Events like solar flares and coronal mass ejections) must be very massive.

39 cm thick (35 g/cm2) polyethylene radiation shield wrapes around 2 centrifuges

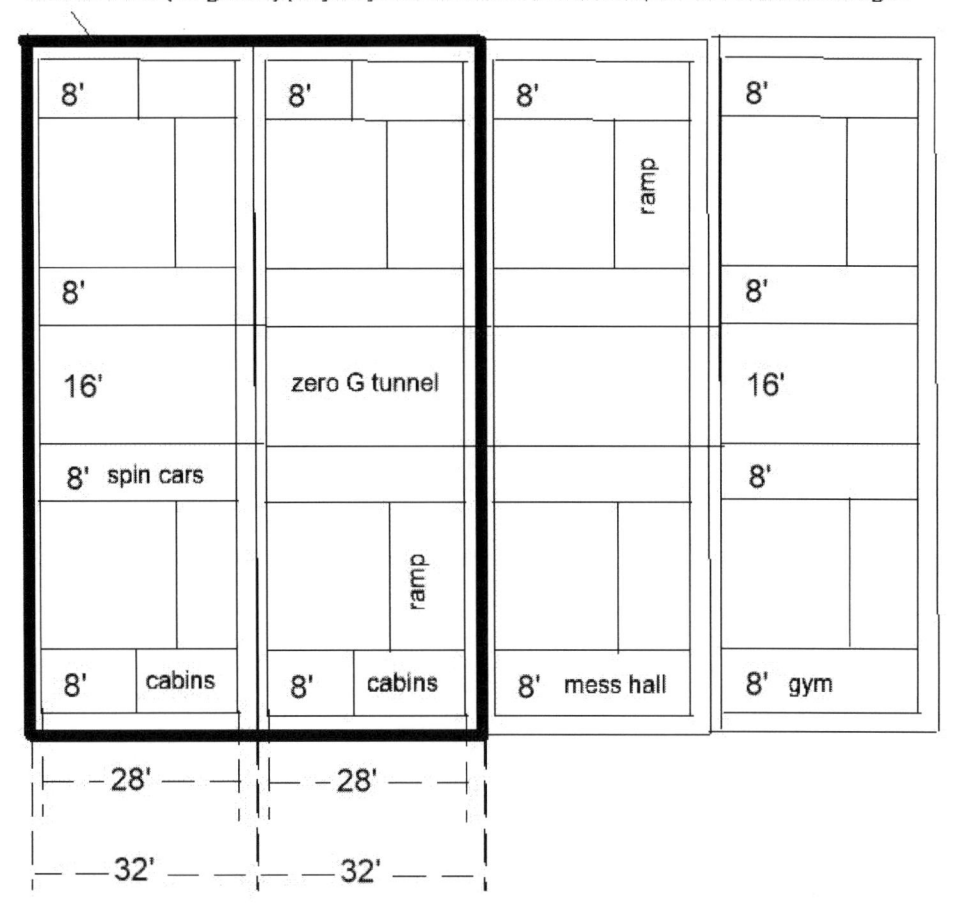

8'	8'	8'	8'
		ramp	
8'			8'
16'	zero G tunnel		16'
8' spin cars			8'
	ramp		
8' cabins	8' cabins	8' mess hall	8' gym

|— 28' —| |— 28' —|
|—— 32' ——|—— 32' ——|

Side view cutaway of centrifuges and shield

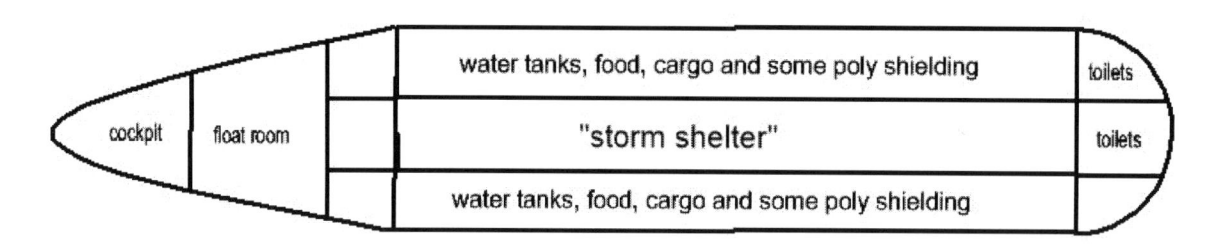

Forward hull of Mars ship. Centrifuges are not shielded. Rear hull contains more cargo and machinery, LSS, etc.

To reduce mass and achieve higher velocity for reduced travel time, Mars ship uses less shielding and water tanks, food, cargo serve as shielding to protect from SPEs. By not shielding centrifuges over 1000 metric tons of mass can be eliminated.

Appendix 2: Silane and Hydrogen Conservation

Imagine 100 metric ton ships that use nuclear thermal rocket engines, a risk that might be acceptable if the ships never enter the Earth's atmosphere, with a specific impulse of 1000 seconds and the delta V from LEO to L1 is 4100 meters per second then 52 tons of LH_2 will be required for one leg of the flight. If chemical propulsion using LH_2 and LOX is used, 153 tons of propellant will be needed. If silane and LOX with a specific impulse of 340 seconds is used then 242 tons of propellant are required. *The interesting thing about this is that chemical propulsion will use more propellant overall but less hydrogen than nuclear propulsion will.* With LH_2 and LOX, an Isp of 450 seconds, and a six to one oxidizer to fuel ratio, 22 tons of hydrogen will be necessary. With silane only ten tons of hydrogen, less than half as much, is needed as with hydrogen and LOX and only a fifth as much as is needed with nuclear.

Propellant Demands for 100 tons ship and a delta V of 4100 m/s

Nuclear rocket Isp 1000 sec. Exhaust velocity = 0.0098(1000) = 9.8 km/sec
e^(4.1/9.8) = 1.52

152/100 = 1.52 152-100= 52 tons LH_2

LH_2/LOX rocket Isp 450 sec. Exhaust velocity = 4.41 km/sec. e^(4.1/4.41) = 2.53   253/100=2.53

253-100= 153 tons propellant using a 1:6 fuel/oxidizer mixture 153/7 = 22 tons LH_2

SiH_4/LOX rocket Isp 340 sec. Exhaust velocity = 3.332 km/sec

e^(4.1/3.332) = 3.42   342/100=3.42

342-100=242 tons propellant

$SiH_4 + 2O_2 ==> SiO_2 + 2H_2O$ Molar masses are: Si = 28 $2H_2$ = 4 $2O_2$= 64

28+4+64=96   4/96 X 242 = 10.08 tons hydrogen

It is likely that hydrogen will be the "pinch point." Silicon and oxygen are abundant in Moon rocks, regolith, and C and S-type asteroids. There is enough solar wind implanted hydrogen on the Moon to provide water for early Moon bases but not enough to fuel rockets. There are huge amounts of hydrogen in polar ices but what will it cost to mine that ice? And do we want to waste such a precious resource that could be of immense value to future lunar civilization? It seems the smart thing to do would be to use less powerful silane and LOX to conserve hydrogen. This also eliminates nuclear dangers. Even greater efficiency might be had if silane is used as a carrier fluid for metal powder fuels (aluminum, magnesium, calcium and/or ferrosilicon) in bipropellant rockets. A great deal of experimentation would be needed to determine the best mixtures. With hydrogen "piggybacked" to Earth orbit with passenger flights and hydrogen from asteroids, lunar hydrogen resources can be conserved for the benefit of future generations.

Appendix 3: Moon Ice

In an earlier chapter we determined that an interlunar ship carrying 200 passengers would weigh about 570 metric tons and use 1984 tons of propellant for one way flight to the Moon or from lunar orbit back to Earth orbit. That would be 661 tons of silane requiring 83 tons of hydrogen. It is thought that there might be five gallons of water ice per cubic yard of regolith in polar craters. That would be about 41.5 lbs. per cubic yard or roughly 20 kilograms per cubic meter. A cubic meter would contain about 2.2 kg. of hydrogen. Since we need 83,000 kg. of hydrogen we would need to mine 37,390 cubic meters of icy regolith on polar crater floors. If we dig to a depth of one meter, we must cover a square area about 193 meters by 193 meters on a side. Since a regulation football field is 120 yards (110 meters) by 53.33 yards (49 meters) that's about seven football fields. That's a lot of mining in the extremely harsh conditions of a dark super cold crater floor. If we could reduce hydrogen demands to one fourth by using silane as a carrier fluid for metal powder fuel the job would be a lot easier. About 20 tons of hydrogen would be needed. Piggybacking liquid hydrogen with passenger flights to LEO might be more practical than polar ice mining.

Appendix 4: Rocket Motor Cooling

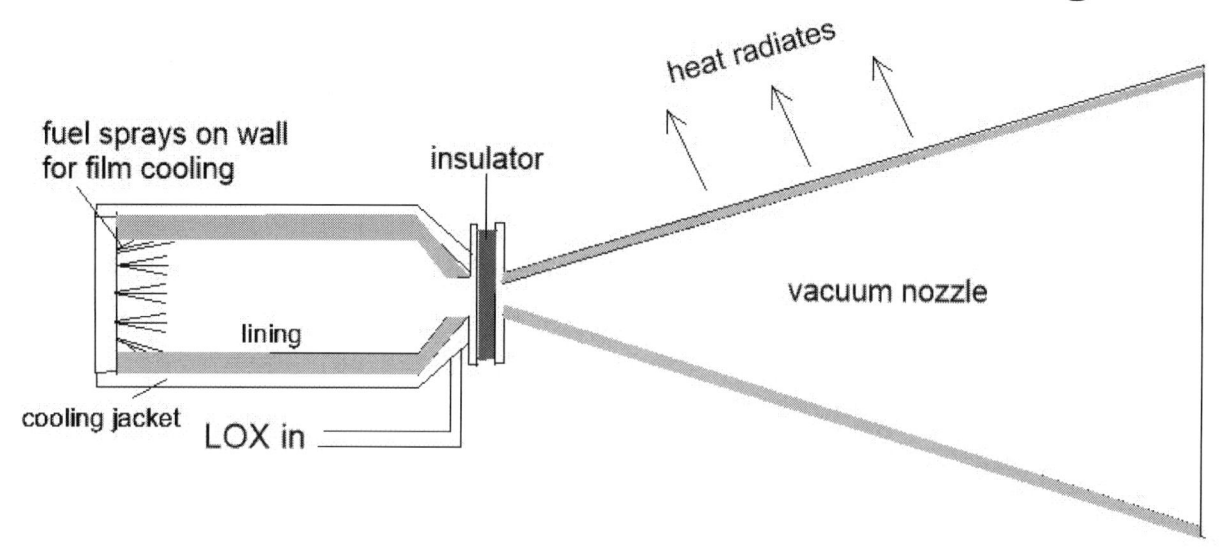

Silane with metal powders suspended in it might clog up narrow cooling passages. It decomposes at about 500° C., so it will not make a good coolant. Liquid oxygen could hot corrode the inside of the cooling jacket; however, experimental rocket motors have used LOX cooling. Perhaps the right metals for the cooling jacket will form a passivation layer that prevents oxidation.

The lining of the combustion chamber will insulate the metal walls of the motor and reduce the amount of heat reaching the cooling jacket. The throat will also need a thick lining and cooling jacket since pressures and temperatures are highest in throat. The lining could be made of lunar cast basalt, silica, alumina, thoria or titania. It would melt and erode and carry away heat. This is called ablative cooling. Silica and metal oxides that form when this fuel burns might fill in the eroded surface of the lining. The nozzle could be disconnected from the firing chamber so that the lining could be serviced.

Fuel would spray on the walls of the chamber and form a film that cools and insulates the metal walls. The vacuum nozzle with a high expansion ratio will improve thrust and exhaust velocity in space. It's size will allow lots of surface area to radiate heat and keep it cool enough not to melt.

The fuel/oxygen mixture ratio might be enriched to reduce temperatures for extended motor lifetime and increase specific impulse and exhaust velocity by adding more low mass hydrogen and unoxidized metals. Low mass particles move faster, therefore they increase exhaust velocity and Isp. The exhaust will contain nanoparticles of SiO_2, H_2O, FeO, CaO, SiO; atoms of Si, Fe, Ca; and molecules of OH and H_2. If Isp increases by enriching fuel/LOX mixture increased hydrogen consumption will not result and efficiency/cost reduction will remain. Tank mass and MR will not be disaffected by dense fuel and performance will be good. Since hydrogen will probably be the most expensive propellant component it is desirable to use less of it and keep fuel costs and ticket prices down.

References

1) Franklin R. Chang-Diaz. "The VASIMR Rocket." Scientific American. Vol. 283 No. 5, Pgs.92-93. November 2000.

2) Andrew V. Iln et al. "VASIMR Human Mission to Mars." Space, Propulsion & Energy Sciences International Forum March 15-17, 2011, University of Maryland, College Park
http://www.adastrarocket.com/Andrew-SPESIF-2011.pdf

3) Al Globus and Joe Strout, Orbital Space Settlement Radiation Shielding, preprint, pg. 14, 2016. http://space.alglobus.net/papers/RadiationPaper.pdf

4) https://www.sciencefocus.com/science/how-strong-does-a-magnetic-field-have-to-be-to-affect-the-human-body/

5) https://www.space.com/23017-weightlessness.html

6)) T.A. Heppenheimer. Colonies in Space. Chp. 5. Stackpole, 1977, 2007
https://space.nss.org/colonies-in-space-chapter-5-first-of-the-great-ships/

7) Wickman Spacecraft and Propulsion Company.
http://wickmanspacecraft.com/lsp/

8) David Dietzler. Mining the Moon: Bootstrapping Space Industry. Chp. 9. Lunar Tourism, 2020.

9) Roger R. Bate et al. Fundamentals of Astrodynamics. Pg 329. Fig. 7.3-1. Dover Publications Inc. New York, 1971.

10) https://www.space.com/21353-space-radiation-mars-mission-threat.html

11) http://apjcn.nhri.org.tw/server/info/books-phds/books/foodfacts/html/data/data2b.html

12) https://www.engineeringtoolbox.com/water-content-d_131.html

13) https://medium.com/@iamgreenified/12-nasa-recommended-air-purifying-plants-that-you-must-have-in-your-house-8797645054b9

14) Michael A. Smith, MD. "CT Scans: How to Protect Yourself from Radiation Exposure." Life Extension Blog. https://blog.lifeextension.com/2010/07/ct-scans-how-to-protect-yourself-from.html

15) https://newatlas.com/nasa-tests-space-gps-pulsars-sextant/52961/

16) MIT Technology Review. An Interplanetary GPS Using Pulsar Signals. https://www.technologyreview.com/2013/05/23/178344/an-interplanetary-gps-using-pulsar-signals/

17) https://optocrypto.com/nasa-tests-celestial-gps-x-ray-positioning/

18) X.P. Deng et. al. Interplanetary Spacecraft Navigation Using Pulsars. National Space Science Center, Chinese Academy of Sciences, Beijing, China. https://arxiv.org/pdf/1307.5375v1.pdf

19) Synthetic Aperture Radar. https://en.wikipedia.org/wiki/Synthetic-aperture_radar

20) 27 Millionaire Statistics: What percentage of Americans are millionaires? (spendmenot.com)

Made in the USA
Columbia, SC
17 November 2021